Control of lead at work
(Third edition)

Control of Lead at Work Regulations 2002

L132

APPROVED CODE OF PRACTICE
AND GUIDANCE

HSE

HSE BOOKS

© Crown copyright 2002

First published as COP2 1980
Second edition 1998
Reprinted 1998
Third edition 2002

ISBN 0 7176 2565 6

The Approved Code of Practice and Guidance

This Code has been approved by the Health and Safety Commission, with the consent of the Secretary of State. It gives practical advice on how to comply with the law. If you follow the advice you will be doing enough to comply with the law in respect of those specific matters on which the Code gives advice. You may use alternative methods to those set out in the Code in order to comply with the law.

However, the Code has a special legal status. If you are prosecuted for breach of health and safety law, and it is proved that you did not follow the relevant provisions of the Code, you will need to show that you have complied with the law in some other way or a court will find you at fault.

The Regulations and Approved Code of Practice (ACOP) are accompanied by guidance which does not form part of the ACOP. Following the guidance is not compulsory and you are free to take other action. But if you do follow the guidance you will normally be doing enough to comply with the law. Health and safety inspectors seek to secure compliance with the law and may refer to this guidance as illustrating good practice.

Contents

Notice of Approval

By virtue of section 16(4) of the Health and Safety at Work etc. Act 1974 ('the 1974 Act'), and with the consent of the Secretary of State for Work and Pensions pursuant to section 16(2) of the 1974 Act, the Health and Safety Commission has on 11 November 2002 approved the revised Code of Practice entitled *Control of lead at work* (Third edition, L132, ISBN 0 7176 2565 6).

The revised Code of Practice comes into effect on 6 December 2002.

The Code of Practice gives practical guidance on the Control of Lead at Work Regulations 2002 and, to the extent that they apply to work with lead, on the Workplace (Health, Safety and Welfare) Regulations 1992.

By virtue of section 16(5) of the 1974 Act and with the consent of the Secretary of State under that paragraph, the Health and Safety Commission has withdrawn the Code of Practice *Control of lead at work* (Second edition 1998, COP2, ISBN 0 7176 1506 5) approved by the Commission on 6 March 1998, which ceased to have effect on 21 November 2002, being the date the Control of Lead at Work Regulations 2002 came into force.

Signed

MARK DEMPSEY
Secretary to the Health and Safety Commission

11 November 2002

Preface

This publication contains the Approved Code of Practice (ACOP) *Control of lead at work*. The ACOP supports the Control of Lead at Work Regulations 2002 (CLAW)(SI 2002/2676). These Regulations contain the provisions of two sets of earlier Regulations (all now revoked): the Control of Lead at Work Regulations 1980 and 1998. In this publication, the text of the CLAW Regulations 2002 is set out in *italic* type, the ACOP text is set out in **bold** type and the guidance text is in normal type.

What is this guidance for?

1 Lead, including its compounds, is a substance that has long been known to have the potential to damage health. Excessive exposure can cause lead poisoning. This guidance, together with the Approved Code of Practice (ACOP) in this publication, supports and amplifies the provisions of the Control of Lead at Work Regulations 2002. These provide safeguards against the risks to health from lead.

2 This publication has been prepared following widespread consultation which included representatives of:

(a) the Confederation of British Industry;

(b) employers' associations representing the lead industry;

(c) the Trades Union Congress and individual trade unions; and

(d) other interested organisations and government departments.

3 The Control of Lead at Work (CLAW) Regulations 2002 (SI 2002/2676) came into force on 21 November 2002 and revoked the CLAW Regulations 1998. The new Regulations have been extended by the health requirements of the European Union's Chemical Agents Directive (98/24/EC), but their aims, and that of their supporting ACOP, are to:

(a) protect the health of people at work by preventing or, where this is not reasonably practicable, adequately controlling their exposure to lead; and

(b) monitor the amount of lead that employees absorb so that individuals whose work involves significant exposure (as defined by the Regulations) to lead at work can be taken off such work before their health is affected.

Application of the Regulations

4 The Regulations apply to any type of work activity, eg handling, processing, repairing, maintenance, storage, disposal etc which is liable to expose employees and any other person to lead as defined in regulation 2, ie to:

(a) metallic lead, its alloys, and all its compounds including lead alkyls; and

(b) lead when it is a component of any substance or material.

5 The lead must also be in a form in which it is likely to be:

(a) inhaled, eg lead dust, fume or vapour;

(b) ingested, eg lead powder, dust, paint or paste; or

(c) absorbed through the skin, eg lead alkyls or lead naphthenate.

6 This means that the Regulations do not apply to work with materials or substances containing lead where, because of the nature of the work, lead cannot be inhaled, ingested or absorbed, eg handling finished pottery products which contain lead.

7 The duty the Regulations place on every employer to prevent or adequately control the exposure of employees to lead applies irrespective of the source of that exposure. For example, the exposure to lead may result from

work with lead or lead compounds being carried out by the employer's own employees, or incidental exposure arising from work nearby being carried out with lead or lead compounds by another employer's employees.

8   Parts of the Regulations apply only if employees' exposure to lead is liable to be significant (see the paragraphs below on 'significant' exposure). The main features of the Regulations are described below.

What do employers have to do?

Assess the risk

9   The Regulations require employers to:

(a)   make a suitable and sufficient assessment of the risks to the health of employees created by the work to include whether the exposure of any employees to lead is liable to be significant;

(b)   identify and implement the measures to prevent or adequately control that exposure; and

(c)   record the significant findings of the assessment as soon as is practicable after the assessment is made.

'Significant' exposure

10   An employee's exposure to lead is significant if one of the following three conditions is satisfied:

(a)   exposure exceeds half the occupational exposure limit for lead; or

(b)   there is a substantial risk of the employee ingesting lead; or

(c)   if there is a risk of an employee's skin coming into contact with lead alkyls or any other substance containing lead in a form, eg lead naphthenate, which can also be absorbed through the skin.

11   If exposure is liable to be 'significant', all the regulations will apply, in particular the need to:

(a)   issue employees with protective clothing;

(b)   monitor lead-in-air concentrations; and

(c)   place the employees under medical surveillance.

12   Employers should use the findings from their assessment to:

(a)   decide the measures needed to control exposure to lead adequately; and

(b)   identify any other measures necessary to comply with the Regulations.

Introduce control measures, and carry out air monitoring if exposure is significant

13   Inhalation is one of the main ways lead can enter the body. The Regulations therefore impose duties on employers to take steps to prevent employees inhaling lead dust, fume and vapour (regulation 6).

14   The employer must:

(a)   introduce control measures to ensure that the amount of lead in the air in the breathing zone of any employee does not exceed the appropriate occupational exposure limit (OEL); and

(b)   carry out a regular programme of air monitoring if the assessment shows that the exposure to lead is liable to be significant, eg above the trigger level of half the OEL, to check that the control measures are working effectively and the OEL is not exceeded (regulation 9).

15   There may be some work activities, eg maintenance operations, when effective engineering controls are not reasonably practicable and so a high standard of personal protection is necessary. In these situations, if exposure to lead is also liable to exceed the OEL, the employer should issue employees with suitable respiratory protective equipment.

Ensure high standards of personal hygiene

16   A high standard of personal hygiene plays a crucial role in controlling lead absorption. Suitable and sufficient washing facilities, including showers where appropriate, are required by regulation 21 of the Workplace (Health, Safety and Welfare) Regulations 1992.[1] This is an important requirement for those exposed to lead at work. Also, lead can be easily absorbed through ingestion. To avoid this risk, the CLAW Regulations impose duties on employers to make sure that employees do not eat, drink or smoke in any place which is liable to be contaminated by lead. To emphasise the importance of this aspect of controlling exposure, the Regulations also place a duty on employees not to eat, drink or smoke in places which they believe may be contaminated by lead.

Place employees under medical surveillance if exposure is significant

17   There is not necessarily a strong relationship between the amount of lead the body absorbs and the concentration of lead-in-air. Consequently, employees whose exposure to lead is significant must be placed under medical surveillance. Regular biological monitoring of the level of the lead in their blood or urine (for work with lead alkyls) can detect any absorption of lead before clinical effects become evident. The Regulations contain biological monitoring indicators to help employers evaluate the effectiveness of their control measures in keeping lead in blood or urine levels at acceptable concentrations:

(a)   *Action levels.* These are concentrations of lead in blood set below the appropriate suspension limit (see next paragraph). If these are reached or exceeded, the employer must:

(i)   carry out an urgent investigation to find out why;

(ii)   review control measures; and

(iii)   take steps to reduce the employee's blood-lead concentration below the action level, so far as is reasonably practicable.

(b)   *Suspension levels.* These are concentrations of lead in blood or urine at which employees are normally taken off work which exposes them to lead, to prevent the risk of lead poisoning.

Provide employees with information, instruction and training

18   The Regulations place a duty on employers to provide employees with suitable and sufficient information, instruction and training. The information to be given to employees includes:

(a)   the possible risks to health of exposure to lead;

(b)   details of the appropriate occupational exposure limit for lead, the action level and suspension level;

(c)   the results of the employer's assessment of the work;

(d)   the appropriate precautions and actions they should take to protect themselves and other employees from exposure to lead; and

(e)   the results of any air monitoring and health surveillance that relate to them personally.

19   This information will enable employees to comply with the duties the Regulations place on them. These duties include making full and proper use of measures to control exposure to lead. This includes any equipment and the facilities the employer provides for that purpose.

Identify the contents of containers and pipes

20   The Regulations also require that unless containers and pipes used for lead are marked in accordance with the Regulations listed in Schedule 2, the employer ensures that their contents and the nature of those contents are clearly identifiable.

Prepare procedures to deal with accidents, incidents and emergencies

21   Under regulation 12, the employer may need to prepare procedures over and above those required by regulation 8 of the Management of Health and Safety at Work Regulations 1999[2] which can be put into effect to deal with accidents, incidents and emergencies relating to the presence of lead at the workplace. However, the employer need not comply with regulation 12 if the quantity, type or form of the lead or lead compounds in the workplace presents only a slight risk to the health of employees, and the employer's measures are sufficient to control the risk.

Outline of the Regulations

22   A diagrammatic presentation of the main provisions of the Regulations is shown in Figure 1.

Consulting employees and/or safety representatives

23   Proper consultation with those who do the work is crucial in helping to raise awareness of the importance of health and safety. It can make a significant contribution to creating and maintaining a safe and healthy working environment and an effective health and safety culture. In turn, this can benefit the business by making it more efficient by reducing the number of accidents and the incidents of work-related ill health.

24   Employers must consult safety representatives appointed by recognised trade unions under the Safety Representatives and Safety Committees Regulations 1977.[3] Employees who are not covered by such representatives must be consulted, either directly, or indirectly through elected representatives of employee safety under the Health and Safety (Consultation with Employees) Regulations 1996.[4] More information on an employer's duties under these Regulations is contained in HSE's free leaflet *Consulting employees on health and safety: A guide to the law*.[5]

Summary of changes to the Regulations

25   The main changes made by the CLAW Regulations 2002 result from the requirement to implement the EC Directive 98/24/EC 'Protection of the health and safety of workers from the risks related to chemical agents at work'.

26   The main changes made by the 2002 Regulations are:

(a)   new definitions: 'hazard'; 'public road'; 'risk'; 'safety data sheet'; 'substance hazardous to health'; 'the risk assessment'; and 'workplace' have been inserted in regulation 2;

(b)   the definitions of 'biological monitoring' and 'medical surveillance' have been amended in regulation 2;

(c)   a reference to a new regulation 12 dealing with accidents has been inserted into regulation 3(1)(b);

(d)   regulation 5 on assessment has been substantially revised to require that:

   (i)   a suitable and sufficient assessment of the risks created by work with lead and the steps needed to meet the requirements of the Regulations is made (regulation 5(1)(a));

   (ii)   the steps identified by the assessment to meet the requirements of the Regulations are implemented (regulation 5(1)(b));

   (iii)   the assessment is to consider a specific list of items (regulation 5(2));

   (iv)   the assessment is reviewed if the results of any monitoring show it to be necessary (regulation 5(3)(c)), or the blood-lead concentration of any employee under medical surveillance equals or exceeds the action level (regulation 5(3)(d));

   (v)   employers who employ five or more employees to record the significant findings of the risk assessment as soon as is practicable after the risk assessment is made and the steps taken to meet the requirements of regulation 6 (regulation 5(4)).

(e)   regulation 6 has been substantially extended:

   (i)   a specific requirement to prevent exposure to lead by substituting a substance or process which eliminates or reduces the risk to the health of employees has been inserted in regulation 6(2);

   (ii)   a list of control measures to be applied in order of priority has been inserted in regulation 6(3);

   (iii)   a requirement to identify the reasons and to take immediate steps when the occupational exposure limit for lead is exceeded has been added to regulation 6(6)(b);

   (iv)   the emergency action the employer should take in the event of the failure of a control measure or an unforeseen event formerly in regulation 6(9) has now been transposed to the new regulation 12;

(f)   new requirements relating to the use of personal protective equipment have been inserted in regulation 8(5), (6) and (7);

(g) a new requirement clarifying how long employers should keep exposure records has been added to the end of regulation 9(4);

(h) a new requirement for employers to keep individual records of monitoring for employees under medical surveillance is in regulation 9(5);

(i) new requirements for the employer to make monitoring records available in certain circumstances are in regulation 9(6);

(j) a further condition on when medical surveillance is appropriate has been inserted at the end of regulation 10(1);

(k) a new requirement allowing an employee access to his personal health record is in regulation 10(6)(a);

(l) a new requirement for the employer to make health records available to HSE is in regulation 10(6)(b);

(m) a new provision which requires an employer who is ceasing to trade to notify and make available to HSE all health records is in regulation 10(6)(c);

(n) new requirements on the employer when medical surveillance reveals an employee's blood-lead concentration to have reached or exceeded the appropriate suspension level are in regulation 10(8)(b), (c), (d) and (e);

(o) regulation 11 on information, instruction and training has been extended to require that:

   (i) it covers the specific list of items at regulation 11(2)(a)-(e);

   (ii) a new requirement on when the information, instruction and training is adapted and how it should be provided is in regulation 11(3);

   (iii) a requirement for the contents of containers and pipes used for lead to be identified is in regulation 11(5);

(p) new requirements to deal with accidents, incidents and emergencies have been inserted into regulation 12;

(q) a new Schedule 2 'Legislation concerned with the labelling of containers and pipes'.

Summary of changes to the ACOP

Changes made to the ACOP

27   The ACOP text gives practical guidance on methods of complying with goal-setting regulations, on what is considered reasonably practicable and on advised but not mandatory methods of meeting legal obligations. A guide to certain requirements in the Workplace (Health, Safety and Welfare) Regulations 1992 and the Management of Health and Safety at Work Regulations 1999 is highlighted by boxes in the left margin.

28   The main changes to the ACOP include:

(a) a complete revision of the paragraphs relating to the assessment, to include guidance on the following specific topics:

(i)   safety data sheets (paragraphs 44-47);

(ii)   the person who carries out the assessment (paragraphs 48-53);

(iii)   making a suitable and sufficient assessment (paragraphs 54-56);

(iv)   exposure to lead and another substance hazardous to health (paragraph 58);

(v)   using personal protective equipment to secure adequate control of exposure (paragraphs 69-72);

(vi)   recording the significant findings (paragraphs 81-87);

(vii)   reviewing the assessment (paragraphs 88-91);

(viii) consulting employees and their representatives (paragraph 92);

(b)   a complete revision of the paragraphs on prevention or control of exposure, to include guidance on the following specific topics:

(i)   prevention of exposure (paragraphs 94-96);

(ii)   control of exposure (paragraphs 97-100);

(iii)   control measures (paragraphs 101-112);

(iv)   action if the OEL is exceeded (paragraphs 115-117);

(v)   suitable respiratory protective equipment, including the fit testing of face-pieces (paragraphs 131-138);

(c)   a complete revision of the structure of the paragraphs relating to the maintenance, examination and testing of control measures (paragraphs 168-210);

(d)   some additional guidance on monitoring exposure, including:

(i)   suitable records (paragraphs 256-259);

(ii)   disposing of records when a business ceases to trade (paragraph 260);

(iii)   access to employees' records (paragraph 261);

(e)   some additional guidance on health surveillance including:

(i)   where a repeat blood-lead test is needed to confirm whether a suspension level has been reached or exceeded, the employer is to make every effort to obtain the result of the repeat test within 10 working days of the initial result becoming available (paragraphs 285 and 297);

(ii)   disposing of records when a business ceases to trade (paragraph 314);

(iii)   access to employees' records (paragraph 315);

(f) a complete revision of the paragraphs relating to information, instruction and training, to include guidance on the following specific topics:

    (i) suitable and sufficient information, instruction and training (paragraphs 322-329);

    (ii) updating information (paragraph 330);

    (iii) making information available to safety representatives (paragraph 331);

    (iv) records of training (paragraph 334);

    (v) providing employees with copies of their records (paragraph 335);

    (vi) people carrying out work on behalf of the employer (paragraphs 337-339);

    (vii) identifying the contents of containers and pipes (paragraphs 340-344);

(g) a completely new section on arrangements to deal with accidents, incidents and emergencies (paragraphs 346-367);

(h) a new 'References' section which lists a number of other related publications.

Figure 1 Outline of the Control of Lead at Work Regulations 2002

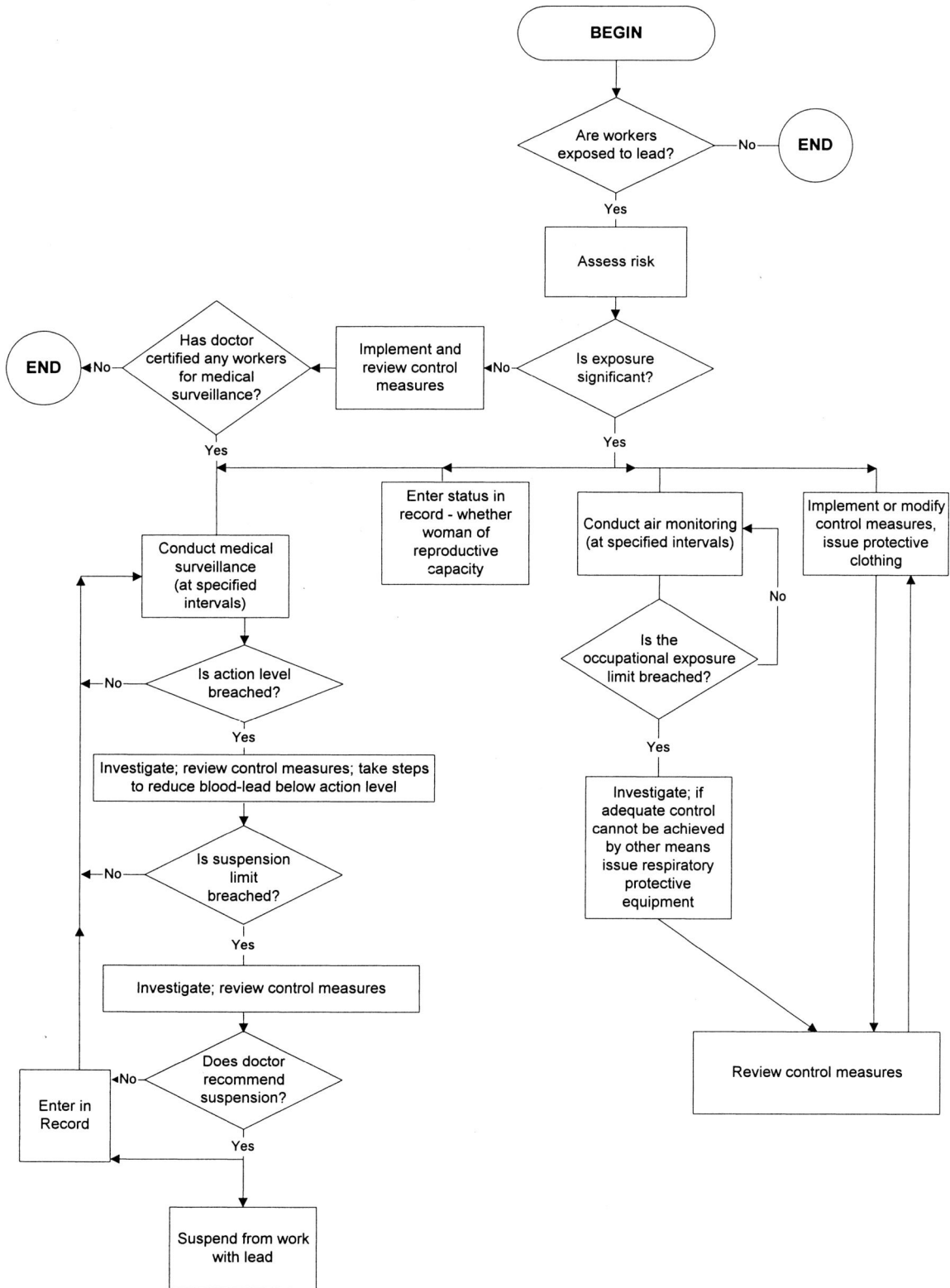

Note: the Regulations also require that, in certain circumstances, employers prepare procedures which they can put into effect to deal with accidents, incidents and emergencies related to the presence of lead at the workplace.

Citation and commencement

These Regulations may be cited as the Control of Lead at Work Regulations 2002 and shall come into force on 21 November 2002.

Interpretation

(1)   In these Regulations -

"action level" means a blood-lead concentration of -

> *(a)   in respect of a woman of reproductive capacity, 25 µg/dl;*

> *(b)   in respect of a young person, 40 µg/dl; or*

> *(c)   in respect of any other employee, 50 µg/dl;*

"appointed doctor" means a registered medical practitioner appointed for the time being in writing by the Executive for the purpose of these Regulations;

"approved" means approved for the time being in writing;

"biological monitoring" includes the measuring of a person's blood-lead concentration or urinary lead concentration by atomic absorption spectroscopy;

"control measure" means a measure taken to reduce exposure to lead (including the provision of systems of work and supervision, the cleaning of workplaces, premises, plant and equipment, the provision and use of engineering controls and personal protective equipment);

"employment medical adviser" means an employment medical adviser appointed under section 56 of the Health and Safety at Work etc. Act 1974;

"glaze" does not include engobe or slip;

"hazard" means the intrinsic property of lead which has the potential to cause harm to the health of a person, and "hazardous" shall be construed accordingly;

"lead" means lead (including lead alkyls, lead alloys, any compounds of lead and lead as a constituent of any substance or material) which is liable to be inhaled, ingested or otherwise absorbed by persons except where it is given off from the exhaust system of a vehicle on a road within the meaning of section 192 of the Road Traffic Act 1988[a];

"lead alkyls" means tetraethyl lead or tetramethyl lead;

"leadless glaze" means a glaze which contains less than 0.5 per cent lead by weight of the element lead calculated with reference to the total weight of the preparation;

"low solubility glaze" means a glaze which does not yield to dilute hydrochloric acid more than 5 per cent of its dry weight of a soluble lead compound when determined in accordance with a method approved by the Health and Safety Commission;

(a) 1988 c.52.

29   An explanation of the definition of a leadless glaze used in these Regulations, and the definition of a low-solubility inorganic lead compound, are set out in Appendix 1.

"medical surveillance" means assessment of the state of health of an employee, as related to exposure to lead, and includes clinical assessment and biological monitoring;

"occupational exposure limit for lead" means in relation to -

(a)   *lead other than lead alkyls, a concentration of lead in the atmosphere to which any employee is exposed of 0.15 mg/m^3; and*

(b)   *lead alkyls, a concentration of lead contained in lead alkyls in the atmosphere to which any employee is exposed of 0.10 mg/m^3,*

assessed -

(i)   *by reference to the content of the element lead in the concentration, and*

(ii)   *in relation to an 8-hour time-weighted average reference period when calculated by a method approved by the Health and Safety Commission;*

30   Occupational exposure limits refer to the concentration in the air to which an employee is exposed. For more information, see paragraphs 212-214.

"personal protective equipment" means all equipment (including clothing) which is intended to be worn or held by a person at work and which protects that person against one or more risks to his health, and any addition or accessory designed to meet that objective;

"public road" means (in England and Wales) a highway maintainable at the public expense within the meaning of section 329 of the Highways Act 1980[a] and (in Scotland) a public road within the meaning assigned to that term by section 151 of the Roads (Scotland) Act 1984[b];

"relevant doctor" means an appointed doctor or an employment medical adviser;

"risk", in relation to the exposure of an employee to lead, means the likelihood that the potential for harm to the health of a person will be attained under the conditions of use and exposure and also the extent of that harm;

"the risk assessment" means the assessment of risk required by regulation 5(1)(a);

"safety data sheet" means a safety data sheet within the meaning of regulation 5 of the Chemicals (Hazard Information and Packaging for Supply) Regulations 2002[c];

"significant" in relation to exposure to lead means exposure in the following circumstances -

(a)   *where any employee is or is liable to be exposed to a concentration of lead in the atmosphere exceeding half the occupational exposure limit for lead;*

(a) 1980 c.66.
(b) 1984 c.54.
(c) SI 2002/1689.

(b)   where there is a substantial risk of any employee ingesting lead; or

(c)   where there is a risk of contact between the skin and lead alkyls or other substances containing lead which can be absorbed through the skin;

31   'Substantial risk' in sub-paragraph (b) above may apply where an employee's hands or face are so liable to be contaminated by lead that there is a high risk that the lead could then be ingested. This could be in circumstances where there is a significant risk of food, drink or smoking materials being contaminated by the transfer of lead from work surfaces, clothing or the employee's skin.

"substance hazardous to health" has the meaning assigned to it in regulation 2(1) of the Control of Substances Hazardous to Health Regulations 2002[a];

"suspension level" means -

(a)   a blood-lead concentration of -

(i)   in respect of a woman of reproductive capacity, 30 μg/dl,

(ii)   in respect of a young person, 50 μg/dl, or

(iii)   in respect of any other employee, 60 μg/dl; or

(b)   a urinary lead concentration of -

(i)   in respect of a woman of reproductive capacity, 25 μg Pb/g creatinine, or

(ii)   in respect of any other employee, 110 μg Pb/g creatinine;

"woman of reproductive capacity" means an employee in respect of whom an entry has been made to that effect in that employee's health record in accordance with regulation 10(14) by a relevant doctor;

(a) SI 2002/2677.

32   See paragraphs 267-269 for guidance on the definition of a 'woman of reproductive capacity'.

"workplace" means any premises or part of premises used for or in connection with work, and includes -

(a)   any place within the premises to which an employee has access while at work; and

(b)   any room, lobby, corridor, staircase, road or other place -

(i)   used as a means of access to or egress from that place of work, or

(ii)   where facilities are provided for use in connection with that place of work,

other than a public road;

33   The definition of 'workplace' is based on that used in the Workplace (Health, Safety and Welfare) Regulations 1992 but is wider in scope as it also applies to domestic premises, ie private dwellings. Certain words in the

13

definition are themselves defined in section 53 of the Health and Safety at Work etc. Act 1974 (HSW Act).

34   In particular, 'premises' means any place (whether or not there is a structure at that place). It includes vehicles, vessels, any land-based or offshore installations, and movable areas to which employees have access while at work and their means of access to and exit from, the workplace. So, common parts of shared buildings, private roads and paths on industrial estates and business parks are included. Public roads which are used to get to or from the workplace are not covered by the definition. However, in some circumstances, a public road may itself become the workplace, and if lead is used or produced during the work activity concerned, CLAW may apply.

"young person" means a person who has not attained the age of 18 and who is not a woman of reproductive capacity.

35   In most cases, female employees under 18 will also be women of reproductive capacity.

(2)   Any reference in these Regulations to either -

(a)   an employee being exposed to lead; or

(b)   any place being contaminated by lead,

is a reference to exposure to or, as the case may be, contamination by lead arising out of or in connection with work at the workplace.

Regulation 3

Duties under these Regulations

(1)   Where a duty is placed by these Regulations on an employer in respect of his employees, he shall, so far as is reasonably practicable, be under a like duty in respect of any other person, whether at work or not, who may be affected by the work carried out by the employer except that the duties of the employer -

(a)   under regulation 10 (medical surveillance) shall not extend to persons who are not his employees other than employees of another employer who are working under the direction of the first-mentioned employer; and

(b)   under regulations 9, 11(1) and (2) and 12 (which relate respectively to monitoring, information and training and dealing with accidents) shall not extend to persons who are not his employees, unless those persons are on the premises where the work is being carried out.

(2)   These Regulations shall apply to a self-employed person as they apply to an employer and an employee and as if that self-employed person were both an employer and an employee, except that regulation 9 (air monitoring) shall not apply to a self-employed person.

(3)   The duties imposed by these Regulations shall not extend to the master or crew of a sea-going ship or to the employer of such persons in relation to the normal shipboard activities of a ship's crew under the direction of the master.

Duty to protect any person likely to be affected by the work

36   An employer who is working with lead, or a substance or material containing it, has a duty so far as is reasonably practicable to protect from

14

exposure to lead anyone else who may be affected by the work. As well as their own employees working with lead, this includes:

(a) other workers, including those employed by another employer, not engaged on work with lead, such as maintenance staff, cleaners etc;

(b) visitors to the worksite;

(c) families of those who are exposed to lead at work and who may be affected by lead carried home unintentionally on clothing and footwear.

37   Other employers of people at the workplace who could be incidentally exposed to lead, such as the maintenance staff, and cleaners referred to in item (a), also have duties under section 2 of the HSW Act and the Management of Health and Safety at Work Regulations 1999 to satisfy themselves that they are taking adequate precautions to protect their employees.

People working under the control and direction of others

38   Although only the courts can give an authoritative interpretation of the law, in considering the application of these Regulations and guidance to people working under another's direction, the following should be considered:

(a) if people working under the control and direction of others are treated as self-employed for tax and national insurance purposes, they may nevertheless be treated as their employees for health and safety purposes;

(b) it may therefore be necessary to take appropriate action to protect them;

(c) if any doubt exists about who is responsible for the health and safety of a worker, this could be clarified and included in the terms of a contract.

39   However, a legal duty under section 3 of the HSW Act cannot be passed on by the means of a contract and there will still be duties towards others under section 3 of the HSW Act. If such workers are employed on the basis that they are responsible for their own health and safety, legal advice should be sought before doing so.

Prohibitions

(1)   No employer shall use a glaze other than a leadless glaze or a low solubility glaze in the manufacture of pottery.

(2)   No employer shall employ a young person or a woman of reproductive capacity in any activity specified in Schedule 1.

40   When considering prospective female employees for work in any of the prohibited activities listed in Schedule 1, employers should take account of the guidance in paragraphs 267-269 in deciding whether a woman is of reproductive capacity.

Assessment of the risk to health created by work involving lead

(1)   An employer shall not carry out work which is liable to expose any employees to lead unless he has -

(a)   made a suitable and sufficient assessment of the risk created by that work to the health of those employees and of the steps that need to be taken to meet the requirements of these Regulations; and

(b)   implemented the steps referred to in sub-paragraph (a).

(2)   The risk assessment shall include consideration of -

(a)   the hazardous properties of the lead;

(b)   information on health effects provided by the supplier, including information contained in any relevant safety data sheet;

(c)   the level, type and duration of exposure;

(d)   the circumstances of the work, including the amount of lead involved;

(e)   activities, such as maintenance, where there is the potential for a high level of exposure;

(f)   any relevant occupational exposure limit, action level and suspension level;

(g)   the effect of preventive and control measures which have been or will be taken in accordance with regulation 6;

(h)   the results of relevant medical surveillance;

(i)   the results of monitoring of exposure in accordance with regulation 9;

(j)   in circumstances where the work will involve exposure to lead and another substance hazardous to health, the risk presented by exposure to those substances in combination;

(k)   whether the exposure of any employee to lead is liable to be significant; and

(l)   such additional information as the employer may need in order to complete the risk assessment.

(3)   The risk assessment shall be reviewed regularly and forthwith if -

(a)   there is reason to suspect that the risk assessment is no longer valid;

(b)   there has been a significant change in the work to which the risk assessment relates;

(c)   the results of any monitoring carried out in accordance with regulation 9 show it to be necessary; or

(d)   the blood-lead concentration of any employee under medical surveillance in accordance with regulation 10 equals or exceeds the action level, *and where, as a result of the review, changes to the risk assessment are required, those changes shall be made.*

> (4) *Where the employer employs 5 or more employees, he shall record -*
>
> (a) *the significant findings of the risk assessment as soon as is practicable after the risk assessment is made; and*
>
> (b) *the steps which he has taken to meet the requirements of regulation 6.*

Purpose of the assessment

41   The purpose of the assessment is to allow employers to make a valid decision about whether the work concerned is likely to result in the exposure of any employees to lead being 'significant' and to identify the measures needed to prevent or adequately control exposure. It also enables the employer to demonstrate readily, both to themselves and to others who may have an interest, eg safety representatives, enforcement authorities etc that:

(a)   they have considered all the factors pertinent to the work;

(b)   they have reached an informed and valid judgement about:

> (i)   the risks the work poses;
>
> (ii)   the steps which need to be taken to achieve and maintain adequate control of exposure;
>
> (iii)   the need for monitoring exposure to lead-in-air at the workplace as part of validating an initial or conditional assessment;
>
> (iv)   the need to place employees under medical surveillance; and
>
> (v)   the need to compile additional procedures over and above those required by regulation 8 of the Management of Health and Safety at Work Regulations 1999 to deal with any accidents, incidents and emergencies involving the lead at the workplace.

42   The assessment may be initial or conditional, ie it may prescribe measures that are likely to adequately control exposure but which need to be tested to confirm their effectiveness. Such measures may subsequently be found to be unnecessary; for example, personal protective equipment (PPE) may be provided as a precautionary measure which is later found not to be required.

43   The CLAW assessment can be made as part of, or as an extension of, the more general risk assessment duties placed on employers by regulation 3 of the Management of Health and Safety at Work Regulations 1999.

Safety data sheets

44   In many circumstances it will only be necessary for employers to read suppliers' safety data sheets to decide whether their existing practices are sufficient to ensure adequate control of exposure. In other more complex circumstances, employers may need to consult HSE guidance notes, manufacturers' standards, technical papers, trade literature and other sources of information to estimate the likely exposure before deciding what control measures they should apply.

45 If a substance or preparation which contains a lead compound is classified as dangerous for supply under the Chemicals (Hazard Information and Packaging for Supply) Regulations 2002 (CHIP), the supplier must provide the recipient with an accompanying safety data sheet (the data sheet), and the employer must consider and take into account the information it provides. It should contain information about the lead compound concerned, which the employer needs in order to ensure that the substance is handled safely in the workplace and the health of employees protected. The data sheet contains information under a number of obligatory headings, eg 'hazard identification', 'exposure controls - personal protection', and so can be a useful source of information in helping employers make the decisions required for the assessment.

46 Accurate, complete and correct information on data sheets is essential when considering workplace controls. If employers have concerns about the quality and reliability of information provided on a data sheet, or if they are unsure of the application of the information to their situation, they should contact the supplier for clarification or for further guidance.

47 The process of carrying out an assessment is not a bureaucratic exercise or simply the collection of information resulting in mountains of paper to be filed away and forgotten. Collecting manufacturer's or supplier's data sheets and other information does *not* in itself meet the CLAW requirements to carry out an assessment. Gathering the information is only the first stage in the assessment process. The information must be used to determine the appropriate control measures to protect the health of employees from any exposure to lead.

The person who carries out the assessment

48 Employers are legally responsible for the assessment and should ensure that the person carrying it out and who provides advice on the prevention and control of exposure is competent to do so in accordance with regulation 11(4). This does not necessarily mean that particular qualifications are required, but the person concerned should have received adequate information, instruction and training (see paragraphs 337-339) and may be, for example, a manager or supervisor with a sound knowledge of health and safety issues, or for more complex situations, an occupational hygienist.

49 Whoever carries out the assessment should:

(a) have adequate knowledge, training and expertise in understanding hazard and risk;

(b) know how the work activity uses or produces lead;

(c) have the ability and the authority to collate all the necessary, relevant information; and

(d) have the knowledge, skills and the experience to make the right decisions about the risks and the precautions that are needed.

50 The person who carries out the assessment does not always have to be fully familiar with and understand the requirements of the CLAW Regulations and their ACOP. However, that person should have access to someone who has a firm grasp of those requirements. This pooling of

knowledge would allow, for example, a supervisor's experience of a process to be combined with the technical and legal knowledge of a health and safety manager.

51   If more than one person contributes to the assessment, the employer should ensure that each person knows precisely what they are to do, and nominate one person to co-ordinate, compile and record the significant findings.

52   Self-employed people should either carry out the assessment themselves, or employ a suitably competent person to do it on their behalf.

53   Employers who decide to use the services of a consultancy to carry out their assessment(s) should ensure that they are competent to do the work. One way to do this is to use one listed in the British Institute of Occupational Hygienists' (BIOH) Directory of Consultancies. You can contact the BIOH for more information at Suite 2, Georgian House, Great Northern Road, Derby DE1 1LT Tel: 01332 298087, website: www.bioh.org.

Making a suitable and sufficient assessment

54   An assessment of the risks created by any work involving exposure to lead should be comprehensive and cover those items listed in regulation 5(2). The assessment should determine whether there is any lead in the workplace to which employees are liable to be exposed, and a form in which it can be inhaled, ingested or absorbed through the skin. Specifically, it should take into account any lead which is:

(a)   brought into the workplace and handled, stored and used for processing;

(b)   produced or given off, eg as fumes, vapour, dust etc by a process or an activity;

(c)   produced from a hidden source, eg from stripping lead-based paint from doors or window frames;

(d)   used for, or arise from maintenance, cleaning and repair work;

(e)   produced at the end of any process, eg wastes, residues, scrap etc;

(f)   produced from activities carried out by another employer's employees in the vicinity.

55   The assessment of the risks created by the work activity should include consideration of:

(a)   where lead is likely to be present and in what form, eg dust, fume or vapour, ie whether it is used or produced and in what amounts and how often;

(b)   the ways in which and the extent to which any groups of people could be exposed, including maintenance workers who may work in circumstances where exposure is foreseeably higher than normal, office staff, night cleaners, security guards, members of the public such as visitors etc, taking into account the type of work and process and any foreseeable deterioration in, or failure of, any control measure;

(c) the need to protect specific groups of employees who may be at particular risk such as young people aged under 18, and women of reproductive capacity;

(d) an estimate of exposure to which employees and any other people affected by the work are likely to be exposed, taking account of:

  (i) the effect of any engineering controls and systems of work used for controlling exposure;

  (ii) any relevant information that may be available about the lead-in-air concentrations likely to be produced by the work concerned; and

  (iii) the effort needed to do the work and how this may affect the rate and volume of air employees breathe (for some activities, employees might breathe three or four times the volume of air that they would breathe at rest);

(e) a comparison of the estimated exposure with the occupational exposure limit for lead;

(f) the records of any blood- or urinary lead absorption available from the biological monitoring of employees similarly exposed to lead, and any record of confirmed lead poisoning;

(g) the potential effects that likely exposure and blood-lead levels may have on the body at different concentrations, seeking the views of the relevant doctor where appropriate; and

(h) whether the work undertaken is likely to result in any employees' exposure to lead being 'significant' as defined in regulation 2, using the guidance given in paragraphs 64 and 65;

(i) how practicable it is to prevent the exposure of employees to lead, eg by using a less hazardous suitable alternative material or a different process. The process of preventing exposure by using an alternative less hazardous substance is a thorough and comprehensive one that should identify suitable alternative substances. It should result in the employer selecting for use the substance that produces the least risk for the circumstances of the work;

(j) if preventing exposure to lead is not possible, the steps which need to be taken to achieve adequate control of exposure in accordance with regulation 6, including if appropriate the selection of suitable respiratory protective equipment; and

(k) systematically identifying the actions necessary to comply with regulations 6-12.

56 Employers should give particular consideration to activities which can give rise to the highest exposures, eg cleaning of equipment, work in confined spaces, or non-routine end-of-shift tasks. Understanding the factors that contribute to employees' exposure will help employers decide how to control it.

Young people

57   Where an employee is a young person, the Management of Health and Safety at Work Regulations 1999 impose a number of additional requirements that the employer must take particular account of when making or reviewing the assessment. These include:

(a)   the inexperience, lack of awareness of risks and immaturity of young people; and

(b)   the extent of the health and safety training provided or to be provided to young people.

Exposure to lead and to another substance hazardous to health

58   Where a work activity may expose employees to lead and to one or more other substances hazardous to health, the employer must consider the possible enhanced harmful effects of combined or sequential exposure. If employees are under medical surveillance which is being supervised by a relevant doctor, the employer should seek advice from the doctor concerned. Otherwise, information may be available from other sources such as the individual suppliers of the substances, trade associations or guidance material etc.

'Significant' exposure to lead

59   As part of the assessment, regulation 5(2)(k) requires the employer to consider whether the exposure of any employee to lead is likely to be 'significant'. In accordance with the definition of that term in regulation 2, the employer should consider all possible routes of exposure to lead, ie inhalation, ingestion and, where appropriate, absorption through the skin. If the employer concludes from the assessment that the exposure of any employee to lead will be or is likely to be 'significant', specific requirements of the Regulations will be triggered, ie the need:

(a)   to provide employees with protective clothing (regulation 6(5));

(b)   to monitor lead-in-air concentrations (regulation 9);

(c)   to place the employees concerned under medical surveillance (regulation 10).

Blood-lead or urinary lead levels which trigger medical surveillance

60   If an employee's blood-lead or urinary lead level is measured as part of the assessment process or at any other time during their employment, and it reveals a level equivalent to or greater than the appropriate level set out in regulation 10(2) and below, the employee should be placed under medical surveillance.

(a)   Blood-lead concentrations:

    (i)   women of reproductive capacity - 20 µg/dl or greater;

    (ii)   all other employees - 35 µg/dl or greater.

(b)   Urinary lead concentrations:

    (i)   women of reproductive capacity - 25 µg Pb/g creatinine or greater;

(ii)   all other employees - 40 μg Pb/g creatinine or greater.

Intermittent exposure to airborne lead

61   An employee's exposure to lead or lead compounds (except lead alkyls) may be intermittent, ie only for a few hours or less during a working week of 40 hours, but that exposure may exceed half the OEL when averaged over 8 hours. In these circumstances, the airborne (inhalation) exposure *may* be regarded as *not* significant for the purposes of medical surveillance where the exposure level:

(a)   is below the OEL when averaged over 8 hours; and *in addition*

(b)   is below half the OEL when averaged over a notional 40-hour week.

62   Employers should take note of the other parts of the definition of 'significant' in regulation 2, and ensure that there is *not* a substantial risk to employees from surface or skin contamination.

Students on work experience programmes

63   In view of the potential hazards of exposure to lead, employers are strongly advised not to allow school students on work experience programmes to do any work where their exposure to lead is liable to be significant.

Work liable to result in significant exposure

64   Some work with lead is liable to result in significant exposure. However, occasional or small-scale activity in the industries and processes concerned may not give significant exposure: the duration, scale and adequacy of the measures to control exposure all have to be considered. Table 1 gives some examples of such work and the sort of industries and processes where significant exposure could be found.

Table 1 Work with lead liable to result in significant exposure

Lead work where there is liable to be significant exposure to lead (unless the employer provides adequate controls)	*Examples of industries and processes where such work could be carried out*
Lead dust and fumes	
1 High-temperature lead work (above 500°C), eg lead smelting, melting, refining, casting and recovery processes, lead burning, welding and cutting.	Lead smelting and refining; casting of certain non-ferrous metals, eg gun-metal; battery grids; leaded steels manufacture; scrap metal and wire-patenting processes, burning of lead-coated and painted plant and surfaces in demolition work; ship-building, breaking and repairing; chemical industry; radiator repair.
2 Work with lead compounds which gives rise to lead dust in air, eg any work activity involving a wide variety of lead compounds.	Manufacture of lead-acid batteries, paints and colours, lead compounds, rubber products; fire assay, ie the use of lead oxides for the assay of precious metals by the process of cupellation; certain mixing and

Lead work where there is liable to be significant exposure to lead (unless the employer provides adequate controls)		*Examples of industries and processes where such work could be carried out*
		melting processes in the glass industry, certain colour preparations and glazing processes in the pottery industry. High-speed mixing and blending of plastics moulding powders containing lead stabilisers or colours. Work with low-solubility lead compounds (as defined in Appendix 1) where poor working practices and standards of cleaning exist (see item 5).
		Battery breaking.
		Manufacture of detonators (explosives industry).
3	Abrasion of lead giving rise to lead dust in air, eg dry discing, grinding, cutting by power tools.	Miscellaneous industries, eg motor vehicle body manufacture and repair of leaded car bodies.
		Firing small firearms on indoor ranges.
		Blast removal and burning of old lead paint.
4	Spraying of lead paint and lead compounds and low-solubility lead compounds (as defined in Appendix 1).	Painting bridges, buildings etc with lead paint.
5	Work with low-solubility inorganic lead compounds (when tested for low solubility as defined in Appendix 1).	Work which is poorly controlled. This might be because of poor ventilation, housekeeping, personal hygiene or lack of proper welfare, eating, drinking or smoking facilities.
6	Paint stripping.	Furniture and joinery restoration, eg removal of old lead paint from antique furniture, doors, window frames etc by immersion in a bath of caustic soda or dichloromethane, and scraping off the residual sludge. May be followed by pressure washing and sanding.
7	Craft work.	Sculpture of bas relief in lead sheet.

Lead alkyls

1	Production of concentrated lead alkyls.	Lead alkyl manufacture.
2	Inspection, cleaning and maintenance work inside tanks which have contained leaded gasoline, eg road, rail and sea tankers and fixed storage tanks.	Oil refineries, oil transport terminals and certain works where tank cars are inspected or repaired.

23

Work not liable to result in significant exposure

65   Some work with lead is not liable to result in significant exposure unless the lead content and/or character of the lead is changed by the work itself, eg where, although the lead content is low and in a finely divided state, it may be concentrated during processing. Table 2 gives some examples of such work and the sort of industries and processes where it could be found.

Table 2 Work with lead not liable to result in significant exposure

Lead work where there is not liable to be significant exposure to lead	Examples of industries and processes where such work could be carried out
Lead dust and fumes	
1 Work with galena (lead sulphide).	Mining and working of galena when its character or composition is not changed.
2 Low-temperature melting of lead (below 500°C). (Such low temperatures control the fume but some care is still required in controlling any dust from dross.)	Plumbing; soldering.
3 Work with materials which contain less than 1% total lead.	
4 Work with lead in emulsion or paste form where the moisture content is such and is maintained so that lead dust and fume cannot be given off throughout the duration of the work.	Brush painting with lead paint and using some stabilisers for plastics.
5 Handling of clean solid metallic lead, eg ingots, pipes, sheets etc.	Miscellaneous metal industries, stockholding, general plumbing with sheet lead.
Lead alkyls	
1 Any exposure to lead alkyl vapours from leaded gasoline where the lead content is limited under the Motor Fuel (Composition and Content) Regulations 1994 (SI No 2295)	Work with such leaded gasoline including, for example, the filling of petrol vehicles on garage forecourts (except for work inside tanks which have contained leaded gasoline as listed in Table 1).

66   The guidance given in Tables 1 and 2 is not exhaustive and the activities listed are not sharply divided. The exposure will vary in different work situations and intensive work or poor control of the activities in Table 2 may give rise to significant exposure.

67   When making an initial assessment of the work that is liable to expose employees to lead, the employer may need to carry out an air monitoring

survey (see regulation 9). However, employers need not carry out this initial survey when:

(a)   exposure to lead can be accepted as significant or insignificant using the information given in Tables 1 and 2, taking the scale and duration of the activity into account for both tables;

(b)   enough reliable data on lead-in-air concentrations are already available for the process or work activity concerned which allows an adequate assessment to be made.

68   The employer and self-employed people working directly with lead should consider the exposure of all those who may be affected by the work, including those listed in paragraph 36.

Using personal protective equipment to secure adequate control of exposure

69   In deciding what measures are needed to control exposure, employers should use personal protective equipment (PPE) where adequate control cannot be achieved solely by application of operational or engineering measures, appropriate to the activity and consistent with the risk assessment. That is, employers may use PPE as secondary protection in combination with other control methods such as local exhaust ventilation, if those other control measures do not adequately control exposure by themselves. Additionally, there may be circumstances where an employer considers it prudent to issue PPE such as clothing, face shields, gloves, aprons etc, not because other control measures are inadequate on their own, but to provide employees with additional protection should any of those measures fail. Note, however, that, in accordance with regulation 6(5), the employer must always provide the employee with suitable protective clothing if the exposure of the employee to lead is liable to be significant.

70   If the assessment shows that control of exposure to lead is likely to be inadequate or become inadequate, the employer should review the effectiveness of all the control measures to decide the steps or, in the case of existing work, the further steps to be taken to obtain and maintain adequate control. As required by regulation 6(3)(c), the employer may have to consider the selection and use of suitable respiratory protective equipment (RPE), but this should only be considered as a method of control after all other measures have been taken.

71   In assessing the likely exposure of employees who have to wear RPE, employers may take into account the additional protection that RPE can provide in reducing the lead-in-air concentration breathed by the employee to the level set by the OEL. As part of the assessment carried out under regulation 5, the employer should consider whether there is a need to provide any RPE for the work and, if so, select a suitable type. The employer's considerations should include the assigned protection factor given to equipment by the manufacturer, and the extent to which other control measures are likely to reduce employees' exposure to or below the OEL.

72   However, employers should be aware that assigned protection factors do not necessarily provide an accurate and reliable guide to the effectiveness that RPE is likely to achieve at the workplace. This is because some RPE wearers

may not use the equipment as they should, or in accordance with the manufacturer's instructions etc, and so the equipment might fail to achieve the level of protection indicated by the equipment's assigned protection factor. Therefore, in selecting suitable RPE, employers should build a safety margin into their considerations to compensate for this uncertainty. They should select RPE capable of providing and achieving a level of protection greater than that estimated to be needed for the work concerned so as to ensure that an employee's exposure to lead-in-air is at least reduced to the level of the OEL.

Outdoor workers at non-permanent workplaces

73   Employers may also need to make an assessment for people who do not work in permanent workplaces, who work outdoors, and whose exposure to lead may vary in its extent from day to day. Their work could include:

(a)   construction, maintenance or demolition of buildings, bridges and other structures;

(b)   installation or maintenance of electricity, water, telephone or railway systems.

74   These activities may give rise to lead exposure through lead burning, paint removal, soldering and handling metallic materials.

75   Although such work with lead may be brief and carried out in the open air, it could still result in exposure to a very high concentration of lead dust or fume before it has had the chance to disperse, for example:

(a)   welding or cutting lead-painted or lead-containing materials carried out close to an employee's breathing zone; or

(b)   lead-burning activities carried out during demolition work, especially in a confined area within a building or structure;

(c)   removal of old lead paint by chipping or by wet or dry shot-blasting.

76   When such activities are planned, it is an essential part of the assessment to identify whether lead paint or other lead-containing material is present. If necessary, a sample of the paint or material concerned should be analysed to find out if it contains lead. If the activities listed in paragraph 75 are carried out on lead-painted or lead-containing materials, the resulting exposure to lead should be assessed as significant because exposure to lead is liable to exceed the OEL.

77   If a small number of people are employed on an outdoor short-term job, the employer still has a duty to provide effective control of exposure to lead dust or fume, although it may not be reasonably practicable to achieve this by engineering or technical control measures. The various factors that can affect this include:

(a)   the type of work concerned;

(b)   the length of time it will last; and

(c)   where it is being carried out.

78   For these work situations, employers may have to rely more on PPE and medical surveillance. Employers will need to take into account the standard of

the hygiene facilities provided on site and the difficulty of providing adequate supervision when assessing the likely exposure of the workers to lead. Since the Regulations also impose a duty to protect from exposure to lead other people who may be affected by the work, employers also need to consider this when planning the precautions and controls for any particular job.

79   The OEL for lead is an 8-hour time-weighted average (TWA) and the assessment of a person's exposure should be made on a day when exposure is likely to be at a maximum level. The reference in paragraph 61 to an intermittent exposure relates only to circumstances when exposure is for a few hours a week. Exposure to lead should be regarded as significant and a worker should be placed under medical surveillance if they are exposed, by inhalation, ingestion or skin absorption to a significant amount of lead for a period of more than a few hours. This is to make sure that a worker's health is not put at risk in circumstances where it is difficult to provide, maintain and supervise control measures.

80   Employers will need to carry out air monitoring in situations where there is not enough information about the lead-in-air concentrations produced by a particular type of work. However, if outside work is short-term, it is not always practicable to carry out air monitoring for every job. Employers should, however, make a realistic estimate of the likely lead-in-air concentration for each job. Employers should use the results of wide-ranging sample surveys to estimate the lead-in-air concentration of jobs of similar types, making allowances for varying factors such as:

(a)   the extent to which the workplace can be enclosed, ie fully or partially;

(b)   any awkward working positions which could bring the worker's face closer than normal to the lead source;

(c)   the composition and condition of the lead source or lead-bearing material; and

(d)   the time taken to do the job.

Recording the significant findings

81   All employers must carry out an assessment but those employing five or more employees must also record the significant findings. Although employers with fewer than five employees are exempt from this requirement, they are strongly advised to record the significant findings of their assessments as a matter of good practice. Employers can use the recorded findings as evidence to:

(a)   show the enforcing authorities that they have carried out a suitable and sufficient assessment in accordance with regulation 5(1); and

(b)   demonstrate that they have systematically considered all the factors relevant to the work, and put in place measures either to prevent exposure or to achieve and maintain adequate control of exposure.

82   The significant findings of the risk assessment should represent an effective statement of hazards, risks and actions taken to protect the health of employees and anyone else who may be affected by the work. Employers will need to record sufficient detail of the assessment itself so that they can demonstrate to a safety representative or inspector etc that they have carried out a suitable and sufficient assessment. The record

may refer to and rely on other documents and records describing procedures and safeguards.

83   The record may be in writing or recorded by other means, eg electronically as long as it is readily accessible and retrievable at any reasonable time for use by employers in reviews or for examination, eg by an inspector, an environmental health officer or a safety representative.

84   The amount of information employers record will be proportionate to the risks posed by the work. In the simplest and most obvious of cases where a work activity involving exposure to lead poses little or no risk, the employer need only record:

(a)   the form the lead or lead compound takes to which employees are or are likely to be exposed, ie pellets, powder, or dust;

(b)   the measures taken under regulation 6 to adequately control exposure, eg taking account of the information provided by the supplier; and

(c)   a statement that because the lead or substance containing lead poses little or no risk, no further detailed risk assessment is necessary.

85   However, where the work concerned presents more of a risk to health from exposure to lead, the significant findings of the assessment should comprise a more comprehensive record and at least include the appropriate items from the following list:

(a)   the processes or activities in which the lead is used, and how employees may be exposed to it;

(b)   the lead or lead compound to which the employees are liable to be exposed and the form in which it occurs, eg powder, dust, fume, vapour or liquid;

(c)   the significant hazards identified by the assessment arising from the exposure to lead under normal working conditions, and in circumstances of an unforeseen incident, accident or emergency occurring which results in an uncontrolled release of lead dust, fume, vapour etc into the workplace;

(d)   details of the extent to which prevention and substitution of the lead or lead compound or process was considered (see paragraphs 94-96, on the substitution requirement in regulation 6(2));

(e)   identification of the separate groups of employees who are liable to be exposed to lead, including any individuals who may be at particular risk, eg young people under 18 and women of reproductive capacity; and

   (i)   how they are likely to be exposed in their areas of the workplace, eg by inhalation, ingestion or by absorption;

   (ii)   their estimated level and length of exposure for the different work activities, taking into account the control measures in place;

(iii) whether the exposure of any employees is likely to be 'significant' in accordance with the definition in regulation 2; and

(iv) whether it is appropriate to place any identified groups of employees under medical surveillance (regulation 10);

(f) the measures in place to achieve adequate control of exposure and the extent to which they control the risks (this need not duplicate details of measures more fully described in other documents such as standard operating procedures but could refer to them) including the use of any PPE and RPE;

(g) the results of any (preliminary) air monitoring to show or confirm the level of exposure to lead in the various activities concerned; an estimate of the level and duration of exposure in normal working conditions taking into account the control measures being used; and the frequency with which any further air monitoring will be carried out;

(h) where appropriate, the reasons for selecting the particular type(s) of protective clothing and RPE issued to employees;

(i) the conclusions reached on the risks to the health of employees and to any other people who may be affected by the work concerned, taking account of the control measures being used;

(j) when the assessment will be reviewed or where appropriate, the period between successive reviews.

86 This record of the significant findings will also form the basis for a revision of the assessment.

When to record the significant findings

87 The employer should record the significant findings when the assessment is made or as soon as is practicable afterwards. In some circumstances, not all the significant findings will have been determined at the same time; some may be awaiting further information before they can be resolved and it will not be possible to record these until then, eg air monitoring results. In these situations, the employer should complete or update the significant findings as soon as the information becomes available. However, the employer must ensure that while waiting for information to confirm the conclusions drawn from the assessment, a cautious approach is adopted to ensure that employees' exposure to lead is adequately controlled, eg in emergency situations where operational decisions have to be made and remedial action taken immediately, or in circumstances where there is a pilot operation which must be run for a period before being assessed completely.

Reviewing the assessment

88 The record of the assessment should be a living document, which must be revisited to ensure that it is kept up to date. The employer should make arrangements to ensure that the assessment is reviewed regularly. The date of the first review and the length of time between successive reviews will depend on the type of risk, the work and the employer's judgement on the likelihood of changes occurring.

89   The assessment should be reviewed immediately:

(a)   when there is evidence to think that it may no longer be valid, for example from:

   (i)   the results of periodic examinations and tests of engineering controls (regulation 8);

   (ii)   the results of monitoring exposure (regulation 9);

   (iii)   the results of medical surveillance (regulation 10), eg the blood-lead level of any employee reaches or exceeds the appropriate action level (see regulation 5(3)(d)), or it shows that lead is being absorbed at higher levels than the original assessment suggested. In these circumstances, the employer may limit the scope of the review to those employees whose blood-lead levels trigger the appropriate action level or to whom the results of medical surveillance apply;

   (iv)   reports or complaints from supervisors, safety representatives or employees about defects in the control systems; or

(b)   new information on health risks;

(c)   where there is to be or has been a significant change in the circumstances of work, especially one which may affect employees' exposure to lead. For example:

   (i)   the introduction of a substitute for lead or lead compound or the type of lead used or its source;

   (ii)   in plant modification, including engineering controls;

   (iii)   in the process or methods of work which is likely to affect the nature of the hazard, eg raising the temperature of molten lead, or changes in the exhaust ventilation system;

   (iv)   in the volume or rate of production; or

   (v)   a reduction in the workforce without any corresponding reduction in the rate of production and the consequent additional pressures on employees.

90   Where the assessment is changed and control measures changed or adapted to meet the new circumstances, employers must take action to implement any necessary changes identified by the review and record afresh the significant findings.

91   When reviewing the assessment, employers should use the opportunity to look again at their prevention or control measures. In particular, they should:

(a)   reconsider whether it is practicable to prevent exposure to lead by changes to the process or by using a suitable substitute or less hazardous form of a lead compound. This may be possible because of technological developments, or changes in the relationship between costs and substances, equipment used and control measures;

(b) re-examine existing control measures to decide whether they can be improved.

Consulting employees and their representatives

92   Employers should involve their employees and/or their safety representatives where they are appointed in the processes of carrying out and reviewing risk assessments. They are in a good position to know what happens in practice and they will use the controls that the employer introduces. Employers can involve their employees as part of their duties under regulation 11 to provide them with suitable information, instruction and training. Employers should also:

(a) tell employees or their workplace representatives the results of the assessment;

(b) explain how control measures are designed to protect their health from exposure to lead; and

(c) explain how any changes will affect the way the employees do the work in the future.

Prevention or control of exposure to lead

(1)   Every employer shall ensure that the exposure of his employees to lead is either prevented or, where this is not reasonably practicable, adequately controlled.

(2)   In complying with his duty of prevention under paragraph (1), substitution shall by preference be undertaken, whereby the employer shall avoid, so far as is reasonably practicable, the use of lead at the workplace by replacing it with a substance or process which, under the conditions of its use, either eliminates or reduces the risk to the health of his employees.

(3)   Where it is not reasonably practicable to prevent exposure to lead, the employer shall comply with his duty of control under paragraph (1) by applying protection measures appropriate to the activity and consistent with the risk assessment, including, in order of priority -

(a)   the design and use of appropriate work processes, systems and engineering controls and the provision and use of suitable work equipment and materials;

(b)   the control of exposure at source, including adequate ventilation systems and appropriate organizational measures; and

(c)   where adequate control of exposure cannot be achieved by other means, the provision of suitable personal protective equipment in addition to the measures required by sub-paragraphs (a) and (b).

(4)   The measures referred to in paragraph (3) shall include -

(a)   arrangements for the safe handling, storage and transport of lead, and of waste containing lead, at the workplace;

(b)   the adoption of suitable maintenance procedures;

(c)   reducing, to the minimum required for the work concerned -

    (i)   the number of employees subject to exposure,

    (ii)   the level and duration of exposure, and

    (iii)   the quantity of lead present at the workplace; and

(d)   the control of the working environment, including appropriate general ventilation; and

(e)   appropriate hygiene measures including adequate washing facilities.

(5)   Where, notwithstanding the control measures taken in accordance with paragraph (3), the exposure of an employee to lead is, or is liable to be, significant, the employer shall provide that employee with suitable and sufficient protective clothing.

(6)   Without prejudice to the generality of paragraph (1), where there is exposure to lead, control of that exposure shall, so far as the inhalation of lead is concerned, only be treated as being adequate if -

(a)   the occupational exposure limit for lead is not exceeded; or

(b)   where that occupational exposure limit is exceeded, the employer identifies the reasons for the limit being exceeded and takes immediate steps to remedy the situation.

(7)   Personal protective equipment provided by an employer in accordance with this regulation shall be suitable for the purpose and shall -

(a)   comply with any provision in the Personal Protective Equipment Regulations 2002[(a)] which is applicable to that item of personal protective equipment; or

(b)   in the case of respiratory protective equipment, where no provision referred to in sub-paragraph (a) applies, be of a type approved or shall conform to a standard approved, in either case, by the Executive.

(8)   Every employer who provides any control measure, other thing or facility in accordance with these Regulations shall take all reasonable steps to ensure that it is properly used or applied as the case may be.

(9)   Every employee shall make full and proper use of any control measure, other thing or facility provided in accordance with these Regulations and, where relevant, shall -

(a)   take all reasonable steps to ensure it is returned after use to any accommodation provided for it; and

(b)   if he discovers a defect therein, report it forthwith to his employer.

(a) SI 2002/1144.

(10) In this regulation, "adequate" means adequate having regard only to the nature and degree of exposure to lead and "adequately" shall be construed accordingly.

Preventing and adequately controlling exposure to lead

93   Regulation 6(1) sets out the general duty: the employer must ensure that the exposure of employees to lead by any route (eg inhalation, ingestion, absorption through the skin or contact with the skin) is either prevented or, where this is not reasonably practicable, adequately controlled. In meeting these requirements, the employer should consider and adopt the best practical measures for achieving the overall protection of employees' health.

Prevention of exposure

94   An employer's overriding duty and first priority is to consider how to prevent employees being exposed to lead. Employers who do not first consider prevention of exposure are failing to comply with a fundamental requirement of the Regulations. The duty in regulation 6(2) to prevent exposure should be achieved by measures other than the use of personal protective equipment. Employers can best comply with this requirement by eliminating completely the use or production of lead in the workplace. This might be achieved by:

(a)   changing the method of work so that the operation giving rise to the exposure is no longer necessary; or

(b)   modifying a process to eliminate the production of a hazardous by-product or waste product; or

(c)   where lead is used intentionally, wherever reasonably practicable substituting a lead-free material which presents *no risk* to health.

95   In many workplaces, it will not be possible or practicable to eliminate exposure to lead completely. Therefore, where it is necessary to use lead, an employer should consider whether it is possible to significantly reduce exposure by using an alternative substance or different form of the lead or a different process which, in the circumstances of the work, presents less risk to the health of employees. This might be achieved by changing the form of the lead so that exposure is negligible.

96   The employer will need to take many factors into account when considering whether to use an alternative substance. These include all the harmful properties of any proposed replacement and whether the requirements of the Control of Substances Hazardous to Health Regulations 2002 (COSHH)[6] apply to the substance concerned. The harmful properties of many potential replacement substances may not all be known, and employers should be aware of this in considering alternatives. The ultimate decision should be based on a balance of any new risks they might present against the potential benefits. For example, in seeking a less toxic substitute chemical for a process, the employer's choice of one with lower toxicity but higher flammability might increase the overall risk if the process has an intrinsic fire risk. Therefore, in considering potential substitutes, employers should be aware of the responsibilities they have under the Dangerous Substances and Explosive Atmospheres Regulations 2002.

Control of exposure

97   Where prevention of exposure to lead is not reasonably practicable, employers must comply with the secondary duty in regulation 6(1) to adequately control exposure. To achieve this, employers must first consider, and, where appropriate for the circumstances of the work, apply the measures set out in regulation 6(3), which are in priority order. This means that employers should consider the application of the measures set out in regulation 6(3)(a) in so far as they are appropriate, before considering those in 6(3)(b) and so on, if further action is needed to achieve adequate control of exposure.

98   Adequate control of exposure to lead should be achieved with measures other than the use of PPE, which should only be used as a last resort and then in addition to other control measures.

99   The employer's aim should be to select the most appropriate controls and those controls should be proportionate to the risks. The employer must take account of:

(a)   the hazardous properties of the lead or lead compound;

(b)   the nature and degree of exposure;

(c)   the conditions of work and the specific work circumstances.

100   It will not always be necessary to apply all the controls described in regulation 6(3) and (4) but it will often be necessary to use a combination of them. The employer's aim should be to select those control measures which in practice will work best to protect the health of employees. The employer should give priority to those controls that contain or minimise the release of contaminants. The administrative and procedural options for control are also important elements which the employer should consider.

Control measures

101   Regulation 6(4) provides a list of typical control measures which employers should consider when applying the control measures set out in regulation 6(3). The objective is to use the findings of the assessment to select the controls or combination of controls that are proportionate to the risk and which will achieve adequate control of exposure.

102   Employers can comply with the requirement in regulation 6(4)(c)(ii) "reducing to the minimum required for the work concerned . . . (ii) the level and duration of exposure" by ensuring that exposure complies with the occupational exposure limit.

103   The requirement in regulation 6(4)(c)(iii) "reducing to the minimum required for the work concerned . . . (iii) the quantity of lead present at the workplace" is not intended to prevent employers buying lead in bulk in order to reduce their costs.

104   The control measures covered by regulation 6(3) and (4) which employers may have to use could be any combination of the following:

(a)   totally enclosed process and handling systems;

(b)   plant or processes or systems of work which:

    (i)   keep the production or generation of dust, fume, vapour to a minimum, eg by modifying a process or changing its conditions such as temperature or pressure to reduce emissions; or

    (ii)   contains it within the plant;

    (iii)   reduces or eliminates the need for maintenance staff to go into hazardous areas; and

    (iv)   limits the area contaminated if spills and leaks occur;

(c)   ventilation:

    (i)   partial enclosure, with local exhaust ventilation;

    (ii)   local exhaust ventilation;

    (iii)   sufficient general ventilation;

(d)   reducing to the minimum required for the work:

    (i)   the number of employees exposed and excluding non-essential employees;

    (ii)   the level and duration of exposure;

    (iii)   the quantities of lead or lead compound used or produced;

(e)   regular cleaning of contamination from walls, surfaces etc or their disinfection;

(f)   providing safe storage and disposal of lead and lead waste;

(g)   prohibition of eating, drinking, and smoking in contaminated areas or other activities which may result in the ingestion of lead;

(h)   hygiene measures: adequate facilities for washing, changing and storage of clothing (see paragraph 126), including arrangements for laundering contaminated clothing, and separate accommodation for clothing worn at work, which may become contaminated by work clothing.

105   Ensuring that employees follow good practice at all times can also play a significant role in helping to secure and maintain adequate control of exposure to lead.

106   In particular, where it is not reasonably practicable for employers to use substitute lead-free materials or low-solubility lead compounds as defined in Appendix 1, the measures used to control all possible routes of exposure to lead should include, where appropriate and reasonably practicable, one or more of the following specific measures:

(a)   using lead or lead compounds in emulsion or paste form to prevent or minimise the formation of dust;

(b) using temperature controls to keep the temperature of molten lead below 500°C, the level at which fume emission becomes significant, though the formation of lead oxide and the emission of dust is still possible below this temperature;

(c) containment of lead, lead materials, compounds, fume or dust in totally enclosed plant and in enclosed containers such as drums and bags. The enclosure should not allow the lead to leak out, and when it is necessary to open it, this should be done under exhaust ventilation if reasonably practicable;

(d) if total enclosure is not reasonably practicable, using an effective exhaust ventilation system which should normally consist of:

(i) partial enclosures such as booths which enclose the lead at source and prevent its escape outside the enclosure by using an exhaust draught;

(ii) various types of hoods which are used when it is not reasonably practicable to enclose the source of pollution. The air movement in the hoods should direct the lead dust, fume or vapour and carry it into the exhaust system. To be effective these hoods should be placed as near as practicable to the point of origin of the lead dust, fume or vapour and draw it away from the employee's breathing zone;

(iii) ductwork with an airflow that is adequate to carry away the dust, fume or vapour;

(iv) a dust and/or fume collection unit with any necessary filtration equipment. Preferably, filtered air from the exhaust system should not be returned to the workplace. If it is, the employer's assessment must demonstrate that there is no significant additional risk to employees' health. To help ensure this, employers should:

- use a high efficiency filtration system and provide effective arrangements for general supply ventilation;

- carefully position the means of returning air to the workplace so that it is diluted with fresh air; and

- ensure that the returning air is not directed into employees' work positions.

High efficiency filtration will usually mean using fabric filters more than 99% efficient, with facilities for monitoring filter performance and detecting filter failure;

(v) fans or other air movers of a suitable type for the system which should be placed in the system after the collection and filtration unit so that the unit is kept under negative pressure to keep any escape of lead to a minimum;

(e) wet methods which include:

(i) wetting of lead and lead materials, eg wet grinding and pasting processes. Wet methods should be used for rubbing or scraping down lead-painted surfaces;

(ii) wetting of floors and work benches while work is being carried out, eg certain work with dry lead compounds and pasting processes in the manufacture of batteries.

Wetting should be sufficiently thorough to prevent dust forming, and the wetted materials or surfaces should not be allowed to dry out. This can create dry lead dust which is liable to be hazardous if it becomes airborne. Water sprays should not normally be used to control an airborne dust cloud as they are unlikely to be effective. Wetting methods should not be used where they are liable to be unsafe, eg:

- at furnaces where they could cause an explosion;

- when lead materials containing arsenides or antimonides could on contact with water produce arsine or stibine gases;

(f) providing and maintaining a high standard of cleanliness.

107 Employers need to give special care and attention to the design of plant, work systems and engineering controls so as to eliminate or reduce workers' exposure to lead, eg:

(a) avoid external ledges on plant on which lead dust can settle;

(b) plant surfaces should be smooth and impervious to make cleaning easy, so far as is reasonably practicable;

(c) provide good joints and seals to prevent leakages.

108 When new plant is commissioned, checks should ensure that design specifications are:

(a) adequate for the specific work activity; and

(b) compliance can be achieved under full production operating conditions.

109 Whenever possible, the design of ventilation plant should be carried out by ventilation specialists.

110 Control measures always need to be set in the context of the overall work environment and should aim to reduce exposure to lead through all stages of the work cycle. Employers should pay particular attention to work methods; for example, when exhaust booths are used for weighing out lead material, consider how the material is to be brought to the booth and handled after work has been completed inside the booth. It is not enough to concentrate only on one aspect of the work and to neglect other aspects which may be equally hazardous.

111 Control measures need to meet the requirements of the Regulations and, so far as is reasonably practicable, conform to current best practices.

112 Employers must ensure that whoever provides advice on the prevention or control of exposure is competent to do so in accordance with regulation 11(4). The people who carry out this work should have adequate knowledge, training and expertise, eg in the design of processes, ventilation and PPE and in the human and technical reasons why control measures fail.

Adequate control

113 Adequate control of exposure to lead covers all routes of possible exposure, ie inhalation, absorption through the skin and ingestion. Adequate control of exposure by inhalation will only have been achieved if the exposure of employees to lead in air does not exceed the appropriate OEL set out in regulation 2. However, because biological monitoring measures absorption of lead in the body by all routes of exposure, overall adequate control of exposure will only have been achieved if the concentration of lead in the employee's blood or urine is kept below the appropriate suspension level (see paragraphs 284-291 and 296-303). In addition, for employees exposed to lead and its compounds, except lead alkyls, the employer is required to investigate and take remedial action if the action level is exceeded (see paragraphs 279-283).

Adequate control of exposure by inhalation

114 When the assessment carried out under regulation 5 suggests that exposure to lead is likely to be significant, the employer should be able to show compliance with the duty to achieve adequate control by carrying out a programme of air monitoring as required by regulation 9. This should aim to show that the appropriate OEL is not exceeded during day-to-day operational activities involving work with lead.

Action if the OEL is exceeded

115 An employer must take immediate steps to remedy the situation where the OEL for lead is shown to have been exceeded. The employer's first step should be to consider if there is a visible, obvious reason for the result(s) which exceed the limit, eg the person to whom the result(s) relates may be subject to higher than normally expected exposure in a job that only that person carries out. If it is an isolated result, or one or two results which marginally exceed the OEL, the employer should consider whether they have real significance and indicate a failure to maintain adequate control, or whether they reflect an error in the measurement method. More measurements may be required.

116 If the employer concludes that the air monitoring results do not indicate adequate control of exposure, the further steps to take include:

(a) checking the control measures to ensure that they are working as they should, and for exhaust ventilation etc, that it is performing to design specification;

(b) liaising with managers, safety representatives and employees to establish possible reasons for the rise in the airborne concentration of lead;

(c) considering whether it is necessary to provide the employees who may be exposed to lead with suitable RPE. This should be a temporary measure only until the situation is returned to normal and adequate control of exposure by inhalation is re-established;

(d) devising and implementing a programme of immediate action to reinforce the control measures; and

(e)   taking further air samples to confirm the concentration of lead in the air.

117   The employer should check that any remedial action to tighten control has been effective. If further air monitoring raises doubts as to whether adequate control is being achieved, the employer should review the assessment to decide whether additional and more stringent controls are needed.

Provision of protective clothing

118   Protective clothing should help to achieve adequate control of lead absorption by protecting personal clothing and the body from being contaminated by lead. It also helps prevent the spread of lead by reducing the chance of contaminated clothing being taken home. The type and design of the clothing and the material from which it is made will be governed by the type and amount of lead to which employees are exposed, which the results of the assessment should identify. Where exposure to lead arises together with hazards such as molten metal, corrosives, wet processes or bad weather, the employer should also take these factors into consideration to achieve a balance that gives the best possible protection that is reasonably practicable.

Suitability of protective clothing

119   In deciding whether protective clothing is suitable and adequate, it must:

(a)   be appropriate for the work concerned and the conditions at the place where exposure to lead may occur;

(b)   cover all parts of employees' own clothing and, so far as is reasonably practicable, effectively resist penetration by lead in the form of dust, fume or lead alkyls and protect the employee's own clothing;

(c)   be designed to avoid, so far as is reasonably practicable, the collection, trapping and retention of lead dust and materials contaminated with lead;

(d)   unless disposable, be made from a material which can be easily cleaned and decontaminated;

(e)   take account of ergonomic requirements and the state of health of the people who may wear it;

(f)   fit the wearer correctly, after any necessary adjustments within the range for which it is designed; and

(g)   be determined whether the materials selected reduce the tendency for the dust to collect on the protective clothing and be re-released, in situations where there is liable to be exposure to airborne lead dust.

120   Manufacturers of PPE must ensure that their products comply with the Personal Protective Equipment Regulations 2002.

121 In many workplace situations, normal overalls should give adequate protection as long as they are in a good condition and are kept clean. In some circumstances, employers may also need to issue employees with protective footwear, for example in foundries and smelters to protect the feet from splashes of molten metal and, where lead dust accumulates on the floor, to prevent contaminated footwear being taken into the uncontaminated areas of the workplace and the worker's home.

122 It is not normally necessary to use protective clothing to provide protection against metallic lead absorption. However, the use of gloves is an effective method of preventing soiling. It is essential to protect the skin when working with lead alkyls or lead naphthenate, as the lead in these substances can be absorbed through the skin.

123 If outer protective clothing, eg a boiler suit or overalls, cannot be worn in such a way that it completely protects the clothing worn underneath from being contaminated by lead, then that 'under' clothing should be classified and issued as protective clothing. Similarly, if additional clothing is required, eg for outdoor work in bad weather, and if that clothing is worn on top of the overalls, it too should be classified and issued as protective clothing.

124 Impermeable protective clothing is essential for work with lead alkyls if there is the risk of skin contact. Certain maintenance operations such as entering storage tanks and other vessels which have contained lead alkyls are potentially extremely dangerous even when the tank or container contains only the fumes from lead alkyls. For these operations, the protective clothing provided should give full protection and could comprise, for example, an air-supplied suit and helmet or hood together with suitable footwear.

125 Whenever practicable, protective clothing and footwear should be issued to individual employees and be clearly identified so that it can be easily sorted and returned to the correct user after being cleaned or repaired. For most items of clothing, at least two sets should be provided: one to wear and one for cleaning and/or repair with more sets being issued as appropriate.

Storage of clothing

The Workplace (Health, Safety and Welfare) Regulations 1992

126 Employers have a duty under regulation 23 of the Workplace Regulations to provide suitable and sufficient 'accommodation' (storage) for clothing. When necessary to avoid any risks to health or damage to the clothing concerned, the employer must also provide separate storage for an employee's own clothing and any protective clothing the employee may have to wear at work. This requirement will normally apply for protective clothing worn for work with lead.

127 For storing clothing, employers should provide enough lockers, hangers, hooks or other storage equipment to cope with the clothing of the maximum number of employees who will need to use the storage. Employers should also make adequate provision for any outside contractors or visitors who may be exposed to lead arising from the work activities.

Provision of respiratory protective equipment (RPE)

128 Regulation 6(3) imposes a hierarchy of control measures and indicates that RPE should always be regarded as the final step in the control regime to be taken to protect employees. Employers should also consider whether it is possible to do the job concerned by another method which will not require the use of RPE or, if that is not reasonably practicable, by adopting other, more effective, safeguards. Employers should only provide appropriate RPE and train employees in its use wherever there is a risk to health and safety that cannot be adequately controlled by other means.

Situations where RPE may be necessary

129 Examples where an employer may have to provide employees with RPE include:

(a) where it is not reasonably practicable to achieve adequate control of exposure by process, operational and engineering measures alone. In these situations, exposure should be reduced so far as is reasonably practicable by these measures, and then, *in addition*, suitable RPE should be used to secure adequate control;

(b) where a new or revised assessment indicates that RPE is necessary to safeguard health and safety until such time as adequate control is achieved by other means;

(c) where urgent action is required, eg because of an accident, incident or emergency involving lead, or an unforeseen event such as a control failure, and the only practicable solution in the time available may be to provide employees with RPE (see regulation 12);

(d) during routine maintenance operations. Although exposure occurs regularly during such work, the infrequency and small number of people involved may not justify the need for process control measures.

130 The employer should always consider and evaluate the limitations of RPE and the costs and practical difficulties of ensuring its continued correct use and effectiveness in the context of the particular work situation and the type and level of exposure to lead.

Suitable RPE

131 For each work activity for which it is foreseen that employees will need to wear RPE, the employer should specify the suitable equipment to be worn to make sure that employees are given adequate protection. To be suitable, RPE must be capable of adequately controlling the inhalation exposure using as a guide the equipment's assigned protection factor as listed in HSE's publication *The selection, use and maintenance of respiratory protective equipment.*[7] The selection and provision of suitable RPE should be based on a range of considerations:

(a) the likely or known airborne concentrations of lead-in-air produced by the work activity concerned;

(b) the level of protection claimed by manufacturers for different types of RPE and identification of those types that will provide a greater degree of protection than that required for likely or known exposure;

(c) the type of work to be done; the physical effort required to do it; the length of time the equipment will have to be worn; the requirements for visibility, comfort and the need for employees to communicate with each other;

(d) the different facial characteristics of the RPE wearers to ensure that the equipment fits correctly, and is matched to the wearer. In addition, the equipment must be matched to the job and the environment in which it is to be used. The selection of suitable equipment should be made in full consultation with the wearers. This will help to ensure that the wearers have the most comfortable equipment best suited for them and which is therefore likely to be the most effective in use;

(e) it must be CE-marked if it was manufactured on or after 1 July 1995 to show that it is manufactured to meet minimum legal requirements. However, if RPE was manufactured before 1 July 1995, it must either be 'CE' marked or HSE approved;

(f) it must be correctly used, ie employees should be properly trained in its use and supervised;

(g) it should be regularly cleaned and checked to ensure that it remains effective.

Fit testing of facepieces

132 The performance of RPE with a tight-fitting facepiece (filtering facepieces, half and full-face masks) depends on good contact between the wearer's skin and the face seal of the mask. A good face seal can only be achieved for this type of facepiece if the wearer is clean-shaven in the region of the seal and the facepiece is of the correct size and shape to fit the wearer's face. If spectacles with side arms and other PPE are also worn, they should not interfere with the correct fitting of the facepiece or the face seal. The performance of RPE with a loose-fitting facepiece (eg visors, helmets, hoods, etc) is less dependent on a tight fit on the face, but nevertheless requires the correct size to ensure the wearer achieves an adequate fit and protection.

133 Employers should ensure that the selected facepiece (tight and loose-fitting types) is the right size and can correctly fit each wearer. For a tight-fitting facepiece (filtering facepieces usually known as disposable masks, half and full-face masks) the initial selection should include fit testing to ensure the wearer has the correct device. The test will assess the fit by determining the degree of face seal leakage of a test agent while the RPE user is wearing the facepiece under test. For full-face masks, a suitable quantitative fit test should be used and the pass level fit factor is 2000. For devices such as filtering facepieces and half masks, the pass level fit factor is 100. For these lower performance facepieces, a suitable and validated qualitative method (often called a semi-quantitative test) can be carried out instead. Employers must ensure that whoever carries out the fit testing is competent to do so in accordance with regulation 11(4).

134 Repeat fit testing will be needed when changing to a different model of RPE or a different sized facepiece or if there have been significant changes to the facial characteristics of the individual wearer, eg as a result of significant weight gain or weight loss or due to dentistry. Repeat fit testing will not be required following a change of employer, provided that the same model of RPE continues to be used by the employee.

135 The quantitative fit testing may be carried out using:

(a) a test chamber which uses a salt aerosol or sulphur hexafluoride gas to assess the face seal leakage; or

(b) a portable device at the workplace which uses the particulates in air to access the face seal leakage; or

(c) a portable device at the workplace which measures pressure variations inside the facepieces to assess face seal leakage.

136 Qualitative test methods use bitter or sweet-tasting aerosols. When the tests are carried out, the facepiece wearer will perform simple exercises as indicated by the competent person carrying out the test.

137 When people work at remote sites, the assessment of any RPE that may be needed should take into consideration the worst conditions that are likely to arise. An adequate supply of RPE, including spare sets of equipment, suitable to meet these conditions, should be made available at the site so that any faulty or damaged equipment can be replaced quickly.

138 More information on the selection, including information on assigned protection factors, use and fit testing of RPE is contained in HSE's publications *The selection, use and maintenance of respiratory protective equipment*[7] and *Fit testing of respiratory protective equipment facepieces*.[8]

Adequate storage facilities for PPE

139 Employers should ensure that storage is provided for PPE so that it can be safely stored or kept when it is not in use. The adequacy of the storage will vary according to the quantity, type and its use, eg pegs, (labelled) lockers, shelves or containers etc. The storage should be adequate to protect the PPE from contamination, loss, theft (especially where equipment is stored at remote sites) or damage by, for example, harmful substances, damp or sunlight. Where quantities of PPE are stored, equipment which is ready for use should be clearly segregated from that which is awaiting repair or maintenance. Where PPE becomes contaminated during use, the storage should be separate from any the employer provides for ordinary clothing and equipment. Employers may also have duties under the Workplace (Health, Safety and Welfare) Regulations 1992 to provide storage for PPE, including RPE.

The provision of washing facilities

The Workplace (Health, Safety and Welfare) Regulations 1992

140 Employers have duties under regulation 21 of the Workplace Regulations to provide suitable and sufficient washing facilities, including showers, if required by the nature of the work or for health reasons, at readily accessible places. The ACOP to the Regulations further recommends that employers should make special provisions for any worker with a disability to have access to facilities which are suitable for their use.

141 For work involving exposure to lead, the washing and changing facilities provided should allow employees to meet a high standard of personal hygiene so as to minimise the risk of their ingesting or otherwise absorbing lead. The design of the washing facilities should be related to the type and level of exposure to lead as indicated by the assessment carried out under regulation 5. Where employees are significantly exposed to lead and if washbasins alone would not be adequate, the washing facilities should include showers or baths. For example where work is carried out in dusty conditions which could result in the whole body being contaminated by lead, the provision of showers or baths would be essential.

142 The assessment may show that the employer needs to pay special attention to the location of facilities to prevent the spread of lead contamination from protective clothing to personal clothing and from one facility to another. When necessary, this can be achieved by:

(a)   separating the changing and storage facilities for protective clothing, protective footwear and RPE from that for personal clothing and footwear not worn during working hours. This can be done by having separate rooms or rooms divided into clean (personal clothing) and 'dirty' (protective clothing) areas. Personal clothing should not be allowed in the dirty room or area. Protective clothing should not be allowed in the clean room or area. The separation of a room into clean and dirty areas should only be considered acceptable when the standard provided is equivalent to that which would be achieved by the use of separate rooms;

(b)   where space permits it is strongly recommended that washing and shower facilities should lie between the facility for storing work clothing and protective clothing. This is to allow employees:

(i)   to remove lead-contaminated clothing and footwear in one room or area;

(ii)   then to pass into the washing and bathing area; and

(iii) finally pass into the clean room or area where they can put on clean clothing and footwear.

143 For certain types of work, such as lead work carried out at premises or sites where such work is not regularly done, eg certain tank-cleaning and lead-burning operations, mobile caravan-type facilities of suitable design should be provided. Similar facilities could also be considered for use in premises where space for further buildings is limited.

144 In laying out their facilities, the employer's aims should be to ensure that they can be conveniently used, and to minimise the spread of contamination. To help achieve these objectives, employers should:

(a)   arrange the facilities, so far as is practicable, in a layout that encourages employees to move correctly between them without causing congestion; and

(b)   consider the best and most convenient location for their facilities in relation to other facilities they provide such as clocking-in points, clothing issue rooms, drying rooms, refreshment points etc.

145 Design features are also important in helping to prevent lead contamination. Walls, ceilings, floors, washbasins, baths and showers should have smooth impervious surfaces which can be easily washed or cleaned and which cannot trap lead and dirt in corners and crevices. Smooth painted washable surfaces, ceramic surfaces and stainless steel surfaces are all suitable.

146 Employers should make sure that washing and shower facilities are supplied with:

(a) washbasins with dimensions that allow people to immerse their arms up to the elbow;

(b) a constant supply of running hot and cold or warm water;

(c) soap or other suitable means of cleaning;

(d) nail-brushes;

(e) individual towels or other means of drying.

147 Towels dispensed from roller towel machines are suitable as long as they provide a clean drying surface for each person but employers should not provide communal towels.

148 The number of washbasins, showers or baths provided should allow the maximum number of people expected to use them at any one time to do so without undue delay. Employers should take account of starting and finishing times and the time available for the use of the facilities. For employees significantly exposed to lead at work and expected to use the facilities at any one time, employers should provide at least:

(a) one washbasin for every five employees; or

(b) 600 mm of trough for every five employees; and

(c) one shower or bath for every five employees who may be expected to shower or bath daily at one time;

(d) an adequate number of showers or baths for people such as maintenance staff and cleaners who are likely to do occasional dirty jobs which may expose them to a high degree of contamination.

149 It is recommended that employees wash and/or shower in *warm* rather than hot or very hot water because *warm* water:

(a) reduces the irritant effect that a combination of soap (or other similar cleanser) with hot or very hot water can have on the skin; and

(b) is far less likely to have any damaging effect on the skin's natural protective properties.

The provision of separate washing facilities

The Workplace (Health, Safety and Welfare) Regulations 1992

150   Regulation 21 of the Workplace Regulations also requires that separate washing facilities are provided for men and women, except where a facility is intended to be used by only one person at a time and is fitted with a door which can be secured from the inside.

Work at more remote sites

151   If the workplace is in a remote area or location, there may be difficulties in providing adequate washing facilities. Nevertheless, basic facilities such as washbasins or bowls, water, soap, nail brushes and towels should be provided as near as possible to the workplace so that workers can wash their hands and faces before eating, drinking or smoking. If the work concerned is liable to expose the workers to a high level of lead contamination, showers or baths should be available for workers to use before they go home. In some circumstances the work may be sited close enough to the base for its facilities to be used, but normally the facilities, eg in a portable building, should be sited at a point convenient for the workplace.

152   Employers should ensure that not only are the hygiene measures provided but also that employees are made aware, through information, instruction and training of why, how and when they must be used. Employers should also ensure through appropriate supervision, that employees use the facilities in accordance with agreed procedures.

153   As with all measures taken to control exposure to lead, the degree of protection provided by the hygiene facilities should be related to the type and level of exposure to lead as established by the assessment.

All reasonable steps to ensure that control measures are properly used and applied

154   The employer should establish procedures to ensure that all control measures, including all items of PPE and any other facilities, are properly used or applied, and are not made less effective by other work practices or improper use. Checking these procedures should form part of normal supervisory duties. Guidance on the control measures and engineering controls that should be checked are set out in the guidance under regulation 8. Procedures will vary but they should include:

(a)   visual checks at least once a shift where appropriate;

(b)   supervising employees to ensure that the defined methods of work are being followed;

(c)   a note of when and where control measures are not being properly used and applied;

(d)   ensuring that where more than one item of PPE is being worn, the different items are compatible with each other; and

(e)   prompt remedial action where necessary.

Environmental considerations

155 Where emissions of lead dust, fume or vapour to the general environment may occur, the employer should liaise with and consult the appropriate environmental enforcing authority for advice and guidance on complying with the relevant regulations.

156 The control of lead in the workplace should not be achieved at the expense of the external environment. Releases of lead to the environment are controlled by Part I of the Environmental Protection Act (EPA) 1990 and a number of other regulations and orders relevant to it. The EPA established two pollution control systems:

(a) an air pollution control system enforced by local authorities in England and Wales, and by the Scottish Environment Protection Agency in Scotland (referred to as 'local enforcing authorities'); and

(b) an integrated pollution control system enforced by the Environment Agency in England and Wales and the Scottish Environment Protection Agency in Scotland.

157 All discharges of trade effluents to sewers are controlled under the Water Industry Act 1991, and discharges to receiving waters, ie rivers, lakes and the sea up to three miles from the coast, are approved by the Environment Agency in England and Wales and the Scottish Environment Protection Agency in Scotland under the Water Resources Act 1991. The Environment Agency and the Scottish Environment Protection Agency set conditions for their approvals which are designed to achieve environmental quality objectives (EQO) set out in EC Directives. There is a range of specific EQOs for lead which depend upon the hardness of the receiving water.

Duties of employees

158 Employees should use the control measures in the way they are intended to be used as they have been instructed. In particular they should:

(a) use the control measures provided for materials, plant and processes, eg if an exhaust booth is provided, the work should be done inside the booth where it is under the influence of the exhaust draught;

(b) follow the defined methods of work;

(c) wear any PPE provided, including any RPE, correctly and in accordance with the manufacturer's instructions; pullovers and jumpers and other items of personal clothing should not be worn on top of protective clothing;

(d) store the PPE when it is not in use in the storage facilities provided;

(e) remove any PPE which could cause contamination before entering a canteen, messroom or other suitably designated clean area to eat, drink or smoke;

(f) practise a high standard of personal hygiene, and make proper use of the facilities provided for washing, showering or bathing and for eating and drinking; eg this includes washing the hands and face and scrubbing the fingernails before leaving to eat, drink or smoke, and making use of any shower or bath facilities provided before leaving for home at the end of the day, especially where there is a high risk of the body being contaminated by the work concerned;

(g) report promptly to the appointed person, ie 'foreman', supervisor or safety representative, any defects discovered in any control measure including defined methods of work, device or facility, or any item of PPE, including RPE.

159 Nail-biting and roll-your-own cigarettes are often implicated in high lead absorption. Employees should use nailbrushes particularly carefully, and should be encouraged to give up smoking or advised to change to manufactured cigarettes if they can't give it up. If employees must roll their own cigarettes, they should roll a supply in a clean environment at home.

160 Employees should also use the canteen, messroom or other eating and drinking facilities the employer provides, as well as any facilities provided for the storage of food and drink, which should not be kept or stored in areas contaminated by lead.

Eating, drinking and smoking

(1) Every employer shall ensure, so far as is reasonably practicable, that his employees do not eat, drink or smoke in any place which is, or is liable to be, contaminated by lead.

(2) An employee shall not eat, drink or smoke in any place which he has reason to believe to be contaminated by lead.

(3) Nothing in this regulation shall prevent the provision and use of drinking facilities in a place which is liable to be contaminated by lead provided such facilities are not liable to be contaminated by lead and where they are required for the welfare of employees who are exposed to lead.

Adequate steps to control ingestion

161 The objective of this regulation is to reduce the risk of ingestion of lead by ensuring that employees do not eat, drink or smoke in places which are contaminated or likely to be contaminated by lead arising from work activities. Therefore employers should reduce the risk of employees ingesting lead by ensuring that they are given adequate information on the specific areas of the workplace that might be contaminated by lead and in which they should not eat (including the chewing of gum or tobacco), drink or smoke. This information should be reinforced by employers displaying prominent notices to identify those areas in which employees either may or may not eat, drink or smoke.

Providing rest facilities

The Workplace (Health, Safety and Welfare) Regulations 1992

162 The arrangements that an employer can make to ensure that employees do not risk ingesting lead by eating, drinking or smoking in places liable to be contaminated by lead will vary according to circumstances. However, employers have a duty under regulation 25 of the Workplace Regulations to

provide suitable and sufficient rest facilities, and these include facilities to eat meals where food eaten in the workplace would otherwise be likely to become contaminated.

163   Where exposure to lead is significant, the employer should provide a canteen, messroom or other suitably designated clean work area for use during breaks. Whenever it is reasonably practicable, these facilities should be provided with suitable tables, seats and storage facilities for food and drink, whether or not the food and drink is supplied by the employer or employee.

164   When it is not reasonably practicable for employees to use the main rest facilities for short (tea) breaks, the employer should provide additional drinking facilities. Under these circumstances the employer may provide drinking fountains or vending machines dispensing liquid refreshment in disposable containers. Where these are used, they should be designed, located and controlled to ensure that the water from the drinking fountain, and the drink and containers from the vending machine are protected from any possible contamination by lead. In very hot work areas, employers should actively encourage employees to drink to replace their lost body fluids.

Providing drinking water

The Workplace (Health, Safety and Welfare) Regulations 1992

165   Every employer is also required by regulation 22 of the Workplace Regulations to provide employees with wholesome drinking water at readily accessible and suitable places.

166   To further reduce the possible risk of contamination, employers should make sure that employees remove any clothing contaminated by lead, and wash and scrub fingernails before eating and drinking. To achieve this, washing and changing facilities should be sited next to, or close by, the designated eating and drinking area.

167   Where exposure to lead is not significant, any designated clean area located away from where lead activities are being carried out would serve as acceptable places where employees can eat and drink.

Maintenance, examination and testing of control measures

(1)   Every employer who provides any control measure to meet the requirements of regulation 6 shall ensure that, where relevant, it is maintained in an efficient state, in efficient working order, in good repair and in a clean condition.

(2)   Where engineering controls are provided to meet the requirements of regulation 6, the employer shall ensure that thorough examination and testing of those controls is carried out -

(a)   in the case of local exhaust ventilation plant, at least once every 14 months; and

(b)   in any other case, at suitable intervals.

(3)   Where respiratory protective equipment (other than disposable respiratory protective equipment) is provided to meet the requirements of regulation 6, the employer shall ensure that thorough examination and, where appropriate, testing of that equipment is carried out at suitable intervals.

(4)   Every employer shall keep a suitable record of the examinations and tests carried out in accordance with paragraphs (2) and (3) and of repairs carried out as a result of those examinations and tests, and that record or a suitable summary thereof shall be kept available for at least 5 years from the date on which it was made.

(5)   Every employer shall ensure that personal protective equipment, including protective clothing, is -

(a)   properly stored in a well-defined place;

(b)   checked at suitable intervals; and

(c)   when discovered to be defective, repaired or replaced before further use.

(6)   Personal protective equipment which may be contaminated by lead shall be removed on leaving the working area and kept apart from uncontaminated clothing and equipment.

(7)   The employer shall ensure that the equipment referred to in paragraph (6) is subsequently decontaminated and cleaned or, if necessary, destroyed.

General

168   The objective of this regulation is to ensure that all control measures perform as originally intended, and continue to prevent or adequately control the exposure of employees to lead. This includes putting right, as soon as possible, any defects found in the controls which could result in reduced efficiency, effectiveness or levels of protection for employees.

169   'Maintenance' means any work carried out to sustain the efficiency of control measures, and not just work carried out by maintenance workers. It includes visual checks, inspection, servicing and remedial work on control hardware and, where appropriate, correction of working procedures.

170   Employers should draw up maintenance procedures to suit individual work situations but they should make clear:

(a)   the engineering controls which need to be maintained;

(b)   the work to be carried out and how it is to be done;

(c)   when the work should be done;

(d)   who is to do the work and who is responsible for it; and

(e)   how any defects found should be put right.

171   Control measures to be maintained should include:

(a)   materials, plant and process control measures;

(b)   protective clothing and RPE;

(c)   washing and changing facilities;

(d) arrangements to ensure that employees do not eat, drink or smoke in places which are contaminated or likely to be contaminated by lead;

(e) cleanliness controls; and

(f) controls to prevent the spread of contamination.

172 Checks of engineering controls should include:

(a) visual checks at least once a shift where appropriate to detect obvious defects in the control measures such as leakages from enclosures etc; and

(b) more thorough examinations and tests using instruments where appropriate to assess the efficiency of technical controls such as exhaust ventilation systems.

173 In most circumstances control measures will include defined working procedures. These should be observed regularly to check that they are being followed. They should be reviewed periodically to confirm that they are still appropriate and workable and also to see whether they can be improved.

People competent to carry out maintenance, examination and tests

174 Employers must ensure that whoever carries out maintenance, examination and tests is competent to do so in accordance with regulation 11(4). People carrying out examination and tests on local exhaust ventilation (LEV) plants should have adequate knowledge, training and expertise in examination methods and techniques. The United Kingdom Accreditation Service (UKAS) publication RG4 *Accreditation for the inspection of local exhaust ventilating plant*[9] provides guidance for the selection of competent people to perform the tasks.

175 Employers and employees should give the person carrying out the thorough examination and test all the co-operation needed for the work to be carried out correctly and fully.

Control measures subject to thorough examination and test

Engineering controls

176 In *all* cases engineering control measures provided to control exposure must be thoroughly examined and tested at suitable or specified intervals. This is to ensure that they are continuing to perform as originally intended. The results of each thorough examination and test should be checked with the assessment carried out under regulation 5 and the requirements of regulation 6 with regard to control. Any defects found as a result of the examination or test should be put right as soon as possible or within the time laid down set by the person who carries out the examination.

177 The form of the thorough examination and test will depend on:

(a) the particular engineering control under consideration, ie its inherent reliability in sustaining the level of control over exposure to lead; and

(b) the type and the extent of the risk posed by lead, ie the consequences of deterioration or failure of the control measure.

178 In the case of LEV plant, the thorough examination and test should cover those items listed in paragraph 185. For all other engineering controls, the examination and test should be sufficient, but no more extensive than is necessary to reveal any defect or latent defect.

Suitable intervals for examinations and tests

179 The employer should decide the intervals between checks and thorough examinations and tests of control measures according to the individual work situation.

180 Factors the employer should consider in deciding what is a suitable interval include:

(a) the type of engineering control in use;

(b) the extent and consequences of the risk of the lead exposure that could result if there was a failure or any deterioration in the effectiveness of the control measure concerned;

(c) the age and condition of the engineering control; its record of reliability and effectiveness; and its reliability to maintain its expected level of control;

(d) the timing of any significant changes to the plant or lead process.

181 However, a general visual check of control measures should be carried out at least once a week, though, where feasible, checking and maintenance procedures should be combined with normal production activity and carried out as part of daily routine.

182 Where the control measures are important for preventing sudden or serious effects on people, the inspection needs to be very frequent, and before each use of controls which are used only occasionally. Condition monitoring, eg airflow sensors in extraction ducts, may need to be continuous and linked to alarms, and there may need to be continuous air sensor operation to detect lead and raise an alarm if a pre-set limit is breached. All arrangements such as these should emerge from the assessment.

Records

183 Employers should keep a suitable record in respect of each thorough examination and test. For LEV plant, the record should contain the information listed in paragraph 185. For all other engineering controls, employers should keep similar information, but adapt it so that it is relevant to the type of engineering control concerned.

Local exhaust ventilation

184 The weekly visual check of an exhaust ventilation system should aim to identify obvious defects in the equipment, such as damage, wear, malfunctioning etc which could result in ineffective extraction and leakage of lead dust, fume or vapour from the system.

185 For all LEV plant, whether fixed or portable, the 14-monthly thorough examination and test should be sufficient to ensure that the LEV plant can continue to perform as intended by design and will contribute to the adequate control of exposure. It should include:

(a) a thorough examination, both internally and externally where appropriate, of the condition of all parts of the system, ie:

    (i) exhaust openings;

    (ii) collection hoods or suction points;

    (iii) ductwork;

    (iv) dust collection and filtration units; and

    (v) fans or air movers;

(b) measurements of static pressure at a point immediately behind each exhaust opening, collection hood or suction point when the equipment is simultaneously extracting from all points served;

(c) measurement of air velocities at the plane of openings to enclosures, collection hoods or suction points for which the standard velocities have been specified;

(d) an assessment of whether or not the lead dust, fume or vapour is being effectively controlled at each exhaust opening, collection hood or suction point. This would be complemented by data on airborne lead concentrations obtained by routine air monitoring.

186 Procedures for putting right any faults should include provision for replacement, repair and remedial action within specified time limits which will range from immediate action to action within a few weeks or months, depending on the degree of risk the fault could present.

187 Checking and maintenance procedures should, where practicable, also be combined with the normal production activity, eg visual checks to detect obvious defects such as damaged protective clothing and leakages from enclosures should be carried out by employees as part of their daily work routine. Employers might wish to delegate maintenance to supervisors while giving the responsibility for overall supervision of maintenance procedures to one particular employee.

188 By following the guidance set out in HSE's publication *Maintenance, examination and testing of local exhaust ventilation,*[10] employers can help to ensure that the examination and testing of their LEV systems are carried out in accordance with the requirement in regulation 8(2).

Suitable records

189 For all LEV plant, whether fixed or portable, the examination and test should be sufficient to ensure that the LEV plant can continue to perform as intended by design and will contribute to the adequate control of exposure. A suitable record in respect of each thorough examination and test of LEV should contain at least the following details:

(a)   the name and address of the employer responsible for the plant;

(b)   the identification and location of the LEV plant, name of manufacturer, model number, and the process involving lead;

(c)   the date of the last thorough examination and test;

(d)   the conditions at the time of test; and whether this was normal production or special conditions;

(e)   information about the LEV plant which shows:

    (i)   its intended operating performance for adequately controlling the lead for the purposes of regulation 6. (Note: If there is no information available on this, it indicates a need for a further assessment in accordance with regulation 5 to show compliance with regulation 6);

    (ii)   whether the plant is still achieving the same performance;

    (iii)   if not, the adjustments or repairs needed to achieve that performance;

(f)   the methods used to make a judgement at (e)(ii) and (e)(iii), eg visual, pressure measurements, airflow measurements and comparison of these with design and other specifications, dust lamp, air sampling, tests to check the condition and effectiveness of the filter;

(g)   the date of examination and test;

(h)   the name, job title (eg senior engineer) and employer of the person carrying out the examination and test;

(i)   the signature, or other acceptable means of identifying the person carrying out the examination and test;

(j)   the details of repairs carried out. The details should be completed by employers responsible for the LEV plant. The effectiveness of the repairs should be proved by a re-test.

190   Examples of the details which should be available in respect of the main components of the LEV system (see (e)(i) above) are as follows:

(a)   *enclosures/hoods* - maximum number to be in use at any one time; location or position; static pressure behind each hood or extraction point; face velocity;

(b)   *ducting* - dimensions; transport velocity; volume flow;

(c)   *filter/collector* - specification; volume flow; static pressure at inlet, outlet and across filter;

(d)   *fan or air mover* - specification; volume flow; static pressure at inlet; direction of rotation;

(e) *systems which return exhaust air to the workplace* - filter efficiency; concentration of contaminant in returned air.

Respiratory protective equipment

191 The maintenance, examination and tests should be in accordance with the manufacturer's instructions. Examinations should comprise a thorough visual examination of all parts of the respirator or breathing apparatus, to ensure that all parts are present, correctly fitted, and the equipment is in good working order. In particular the examination should ensure that the straps, facepieces, filters and valves are sound and in good working condition. For powered and power-assisted respirators, tests should:

(a) be made on the condition and efficiency of those parts;

(b) ensure that the battery pack is in good condition; and

(c) ensure that the respirator delivers at least the manufacturer's recommended minimum volume flow rate.

192 For RPE incorporating compressed gas cylinders, tests should include the condition and efficiency of all parts, the pressure in the cylinders and the volume flow rate.

Frequency of examination and tests

193 The quality of the air supplied to a breathing apparatus should be tested at least once every three months and more frequently when the quality of the air supplied cannot be assured. Where the air supply is from mobile compressors, the employer should ensure that wherever a compressor is located, the quality of air it supplies is not compromised by nearby contaminants. In every case, the air supplied to a breathing apparatus should meet the quality standard recommended in clause C.1.2 of BS 4275: 1997 *Guide to implementing an effective respiratory protective device programme.*[11] However, BS 4275 recommends that all contaminant levels should be below one tenth of the OEL. As it is not reasonably practicable to test for all contaminants, the risk assessment made under regulation 5 should guide what other substances may require testing.

194 Thorough maintenance, examinations and, where appropriate, tests of items of RPE, other than one-shift disposable respirators, should be made at least once every month, and more frequently where the health risks and conditions of exposure are particularly severe.

195 However, in situations where respirators are used only occasionally and for short spells, an examination should be made prior to next use and maintenance carried out as appropriate. The person who is responsible for managing the maintenance of RPE should determine suitable intervals between examinations, but in any event, the intervals should not exceed three months. Emergency escape-type RPE which an employer provides should be examined and tested in accordance with the manufacturer's instructions.

196 Suitable arrangements should ensure that no employee uses RPE which has previously been used by another person unless it has been thoroughly washed and cleaned in line with the maker's instructions.

Suitable records

197 The record of each thorough examination and test carried out should include:

(a)   the name and address of the employer responsible for the RPE;

(b)   particulars of the RPE equipment and of the distinguishing number or mark, together with a description sufficient to identify it, and the name of the maker;

(c)   the date of examination and the name and signature or other acceptable means of identifying the person carrying out the examination and test;

(d)   the condition of the equipment and details of any defect found, including (for canister or filter respirators) the state of the canister and the condition of the filter;

(e)   for any self-contained compressed air/gas breathing apparatus, the pressure of air/gas in the supply cylinder;

(f)   for any powered/power-assisted respirators and breathing apparatus, the volume flow rate to ensure that they can deliver at least the manufacturer's minimum recommended flow rate.

Keeping records

198 Employers may keep records in any format, eg on paper or by other means, eg electronically. They must be clear and understandable and be readily accessible and retrievable at any reasonable time for examination by safety representatives, inspectors etc. Records should be held at the premises where the work with lead is carried out. Where this is not possible, eg because of the itinerant nature of the work such as demolition work, the records should be kept at the employer's designated office.

199 All records should allow people other than those who carry out the examination and test of control measures to:

(a)   decide whether the work has been done and completed at the appropriate time;

(b)   learn of any defects or matters requiring attention and of the remedial action taken;

(c)   use the records as a measure of performance to decide whether any necessary improvements need to be made to the controls; and

(d)   provide data for future research with a view to improving controls.

Keeping the workplace clean

200 'Control measure' as defined in regulation 2 includes all measures taken to prevent or reduce exposure to lead. They include keeping the workplace sufficiently clean and the proper use by employees of suitable washing facilities

8

8

56

provided by the employer, because these activities play an important part in helping to achieve overall control of exposure to, and absorption of, lead.

201 The employer's procedures for ensuring that the workplace is sufficiently cleaned should specify:

(a) the name or job status, eg shift 'foreman', of the person responsible for overall supervision of the cleaning procedure;

(b) the area, plant, equipment or clothing to be cleaned;

(c) how often such cleaning should be carried out;

(d) the method(s) of cleaning.

202 Cleaning procedures should concentrate in particular on areas where dust is liable to accumulate and where it can be dislodged and become airborne, eg ledges, shelves, porous surfaces, floors, inside walls, ceilings, workbenches and external surfaces of machinery. Dust should not be swept up but removed by using a vacuum equipped with a high-efficiency particulate arrester (HEPA) filter, or by wet cleaning.

203 Cleaning should be carried out as frequently as is necessary to secure cleanliness and the removal of lead deposits so reducing the risk of inhalation and ingestion of lead. In particular the following should be cleaned at the minimum intervals specified below:

(a) floors and workbenches at least once a day;

(b) RPE at the end of every shift or work period;

(c) protective clothing should be washed whenever contaminated after use, and cleaned or renewed at least once a week; and

(d) washing and changing facilities and facilities for eating and drinking should be washed or cleaned etc at least once a day.

204 The cleaning methods used should not create a risk from lead for cleaners or other people or spread contamination. Suitable methods include:

(a) wet cleaning; and

(b) the use of mobile or fixed vacuum cleaning apparatus equipped with high-efficiency (HEPA) filters.

205 When wet cleaning is carried out, care needs to be taken to make sure that surfaces are kept free from wet sludge which could be transferred to clean areas which, when dry, could create a substantial amount of airborne dust.

206 Employees should be encouraged to keep their immediate work areas clean and to clear up and appropriately dispose of any spillages they may cause, as directed by their employer.

Cleaning of protective clothing

207 Employers should ensure that any protective clothing which is liable to be contaminated with lead does not leave the premises, except to be

cleaned by a specialist laundry. Protective clothing and towels should not be taken home by employees for cleaning, but wherever possible, these should be cleaned at the premises where work with lead is carried out.

208 When laundering is carried out in-house, the employer should make sure that the cleaning arrangements are adequate to prevent the spread of contamination and not put employees who work on the laundering at risk.

209 If contaminated protective clothing and towels are not cleaned on the employer's premises, they should be sent to a specialist laundry equipped to do the work. The laundry should be alerted to the lead risk and of the need to make sure that cleaning is done without any risk to employees. When clothing contaminated by lead is sent to a specialist laundry, it should be placed in suitable impermeable and securely fastened containers, or heavy duty plastic bags, and labelled 'lead-contaminated clothing'.

210 Where wet work is carried out, protective plastic aprons contaminated by lead should be thoroughly washed (leather aprons should be wiped clean with a wet cloth) and hung up to dry before employees leave the work area for a rest or a meal break. This is important to prevent any lead residue on the apron from drying out and later contaminating the wearer when it is reused.

8

Regulation 9

Air monitoring

Regulation

(1)   Where the risk assessment indicates that any of his employees are liable to receive significant exposure to lead, the employer shall ensure that the concentration of lead in air to which his employees are exposed is measured in accordance with a suitable procedure.

(2)   Subject to paragraph (3), the monitoring referred to in paragraph (1) shall be carried out at least every 3 months.

(3)   Except where the exposure referred to in paragraph (1) arises wholly or in part from exposure to lead alkyls, the interval between each occasion of monitoring may be increased to a maximum of 12 months where -

(a)   there has been no material change in the work or the conditions of exposure since the last occasion of monitoring; and

(b)   the lead in air concentration for each group of employees or work area has not exceeded 0.10 mg/m³ on the two previous consecutive occasions on which monitoring was carried out.

(4)   The employer shall ensure that a suitable record of monitoring carried out for the purpose of this regulation is made and maintained and that that record or a suitable summary thereof is kept available for at least 5 years from the date of the last entry made in it.

(5)   Where an employee is required by regulation 10 to be under medical surveillance, an individual record of monitoring carried out in accordance with this regulation shall be made, maintained and kept in respect of that employee.

(6)   The employer shall -

9

(a) on reasonable notice being given, allow an employee access to his personal monitoring record;

(b) provide the Executive with copies of such monitoring records as the Executive may require; and

(c) if he ceases to trade, notify the Executive forthwith in writing and make available to the Executive all monitoring records kept by him.

211 If there is significant exposure to lead, the monitoring of employees' exposure should be carried out by both air sampling and by measuring the concentration of lead in their blood or their urine (for work with lead alkyls). These two approaches have complementary roles.

The occupational exposure limits for lead

212 The OELs relate to an employee's personal exposure to lead-in-air. Therefore, the measurement of exposure will normally involve the collection of a sample of air from the employee's breathing zone using personal sampling equipment and techniques. However, during the manufacture of lead alkyls, sampling may also involve continuous sampling of the workplace atmosphere using static sampling equipment. When continuous static sampling procedures are used, the employer should be able to show that the results are representative of the concentration of lead-in-air in the breathing zone of the employees. This is done by occasionally supplementing the continuous sampling with personal sampling.

213 The method approved by HSC for calculating an occupational exposure over a specified reference period is set out in Part 3 - 'Applying occupational exposure limits: Calculation methods. Calculation of exposure with regard to the specified reference periods' of HSE's annual publication *Occupational exposure limits* (EH40).[12] Some examples are set out in Appendix 3 of this publication.

214 For some work, employees may be exposed to lead or lead compounds, except lead alkyls, intermittently, ie for only a few hours or less during a standard working week of 40 hours. Typically, this may occur where lead is used outside the mainstream sectors of the lead industry, but these employees are still subject to the general control provisions set out in regulation 6 of CLAW. Under these circumstances, the lead-in-air concentration may occasionally exceed the OEL for lead and its compounds when exposure is averaged over the time-weighted reference period of 8 hours set out in the Regulations. For employees whose exposure to lead is limited to the extent described, employers may substitute 40 hours as the base reference period for the OEL for the purpose of deciding whether a concentration of lead-in-air in the workplace exceeds the OEL and there is adequate control of exposure by inhalation.

General requirements for air monitoring

215 Measurements of lead-in-air concentrations to show compliance with the exposure limit should be made in the breathing zone by means of personal sampling equipment. Static sampling alone is not enough, and where it is used, it should only be regarded as a supplement to personal sampling. Static sampling measurements may be useful in checking on the continuing efficiency of engineering control measures.

216 Measuring the concentration of lead in the blood, or in urine where there is exposure to lead alkyls, provides the best indication of the absorption of lead in the body for each employee or for a group of employees. The results of biological testing can thus reflect the effectiveness of hygiene provisions and precautions in general.

Specific requirements for air monitoring

217 For measuring the concentration of lead-in-air, other than for lead alkyls, suitable air monitoring should involve the use of personal air sampling techniques and strategies. Employers must ensure that whoever carries out monitoring of exposure is competent to do so in accordance with regulation 11(4). The person who does the work should have adequate knowledge of the occupational exposure limits, exposure monitoring strategies and training and expertise in measurement techniques and their interpretation.

218 In 1993, the European Standards Organisation (CEN) and other organisations agreed a convention for measuring the inhalable fraction of suspended matter and this has been published in Great Britain as British Standard BS EN 481: 1993 *Workplace atmospheres. Size fraction definitions for measurement of airborne particles.*[13]

219 Suitable air sampling techniques that employers should use to comply with the requirements of the British Standard, together with guidance on analysing the results, are described in full in a publication in HSE's Methods for the Determination of Hazardous Substances (MDHS) series: *Lead and inorganic compounds of lead in air. Laboratory method using flame or electrothermal atomic absorption spectrometry.*[14] The method involves the use of personal sampling techniques for estimating an 8-hour time-weighted average of lead-in-air concentration so that it can be compared with the occupational exposure limit. Appendix 2 lists the technical specifications of the equipment that should be used for monitoring concentrations of lead in air.

Factors that can affect exposure to lead

220 It is not possible to draw meaningful conclusions about typical or long-term exposure patterns to lead based on a single measurement of exposure at one point in time. This is because variations in plant, process, work and other conditions may directly affect the concentration of lead-in-air.

221 For example, a manufacturing process may involve a variety of tasks and a range of exposure conditions. Therefore, designing a monitoring strategy will be greatly influenced by the conditions presented by the process or activity involving lead. The scope of the exposure assessment will depend on the complexity of the process or activity.

222 In designing an air monitoring strategy, the factors to be considered include:

(a)   the physical and chemical properties of the lead dust, fume or vapour, eg the size of the particles;

(b)   the locations of the various sources from which it is released;

(c)   the rate, duration, speed and concentration of release from each source;

(d) the design of the work process and control measures;

(e) variations that may occur in the process and their frequency, eg different tasks may present different exposure conditions;

(f) how any dust, fume or vapour is dispersed into the workplace air by the general or local ventilation and affected by the ambient conditions, eg temperature, humidity etc.

223 Employers should be aware how employees' work can influence the results of air sampling, eg:

(a) the type and position of each source of release of lead dust etc relative to the employee;

(b) the length of time the employee spends in the area where the lead is present;

(c) whether the employee has direct control over the task or process;

(d) the extent of the employee's appreciation of the risks of working with lead and whether that results in good work practices.

Preliminary survey and initial appraisal

224 There may be enough data for planning air monitoring procedures without the need for an initial appraisal or preliminary survey if employers already have comprehensive information on:

(a) the hazards;

(b) the potential risks; and

(c) employee exposures from their various activities involving lead.

225 However, where exposure to lead is significant but there is not enough basic information and lead-in-air data available to provide the basis for drawing up an air monitoring programme, employers should carry out a preliminary assessment and survey.

226 An initial appraisal to identify the risks and potential hazards is important in helping to decide the need and extent of exposure monitoring. Some of the basic questions that will need to be answered are already listed in paragraph 222. Others include:

(a) the processes or operations where exposures are likely to occur;

(b) which groups of employees are most likely to be exposed;

(c) whether protective clothing and/or RPE will be worn and their effectiveness; and

(d) the OEL for the type of lead concerned; and whether exposure is likely to exceed half the limit (which triggers the definition of 'significant' exposure), or the limit itself when employees have to be provided with RPE.

227 This information can be supplemented with the results of other tests to help decide whether the work concerned may pose a risk to employees' health.

For example:

(a) a dust lamp allows very fine particles of dust to be seen which are invisible under normal light. This can help those concerned identify emission sources and watch the movement of airborne dust;

(b) smoke tubes can show the movement of air under the influence of draughts, general and local exhaust ventilation systems and show what the effects are.

228 The data collected during the initial appraisal should provide an accurate description of individual exposure to airborne lead for comparison with the OELs. From this comparison it should be possible to conclude whether the level of exposure by inhalation is not significant as defined by the Regulations and whether it is necessary to carry out routine air monitoring.

229 However, if the results show that exposure is significant, they should be used to decide the number and location of sampling points (both personal and static) and monitoring carried out in accordance with regulation 9(2).

230 Employers will need to conduct a more comprehensive survey if the initial appraisal suggests for example that:

(a) an exposure risk exists but the extent is uncertain;

(b) major changes have been made to the process, procedures or control measures since the last assessment;

(c) non-routine or unusual operations are planned; or

(d) a new process or activity is being introduced.

231 Before carrying out the survey, those employees most likely to be at significant risk of exposure should be identified and the conditions which lead to this exposure addressed (see paragraph 222). The survey undertaken should estimate the personal exposure of the employees but also give an indication of the efficiency of process and engineering controls, especially in 'worst case' situations.

Identifying the employees to be monitored

232 All employees who are exposed significantly to lead, and for whom data are required, should be individually identified. They should include not only those who are continually exposed to lead dust, fume or vapour, eg process operators, but also those who may be exposed to significant concentrations for short intermittent periods such as maintenance workers, cleaners and crane drivers.

233 When a number of employees are working in similar jobs, sampling can be carried out most effectively by dividing the employees into groups doing identical or similar work in well-defined work areas. A work area may be the whole or part of one room. Groups of people doing the same work but in different rooms or buildings should be treated separately. If there is considerable variation between the work of each shift, eg permanent day and night shifts, then each shift should be considered as a separate group. When grouping employees, care should be taken to identify any individuals with unique exposure patterns so that their exposure can be measured separately.

234 When a group measurement is planned, the air sample should be carried out on a random number of individuals from the whole group. Enough samples should be taken to cover the range of activities within the group and provide a sound numerical basis on which to assess the exposure of the group and any future changes in exposure.

235 **The number of people to be sampled in each group during each assessment will vary according to different work circumstances, but should not be less than one for every ten people exposed.**

236 If the group comprises nine or fewer employees, the employer should decide how many employees need to be sampled. The number selected should ensure that the sampling results are valid for all individuals within the group. Sampling should therefore adequately cover all individual activities, work and exposure patterns, including possible variations there may be, such as between day and night shifts.

237 It is important that the results are carefully analysed to make certain that they are valid for all the individuals who make up the group. A wide distribution of results may suggest considerable variation within the group and could mean that individuals have not been grouped correctly. In these circumstances a further sub-division of the group could focus on those at risk and make more efficient use of resources.

Routine monitoring

238 When the risk to employees is being adequately controlled, the data obtained from the initial appraisal or more comprehensive survey showing individual or group exposure to lead, can be used as the basis for preparing a routine air monitoring programme designed to monitor the effectiveness of control measures.

239 Routine sampling should be planned and carried out in a similar way to the preliminary survey, ie random coverage of groups and full shift coverage of work. However, if the preliminary survey or other results show there is a consistent personal exposure to lead over a whole work period, the length of sampling time during routine sampling can be reduced as long as it remains representative.

240 Enough air samples should be taken during routine monitoring to identify any major changes in exposure levels. Any abnormal results, especially any above the OEL, should be carefully investigated to find out the reason for the abnormality and to make sure that the employees concerned are adequately protected until further controls are introduced. Employees and their representatives should be told as quickly as possible of any results above the OEL and the reason why these results occurred. They should also be told of any additional measures which have been taken to reduce exposure and to prevent a repeat of the incident.

241 If lead-in-air concentrations are significant, air monitoring for each group of employees or each work area should be carried out at least once every three months. The frequency can be reduced to once every 12 months if the conditions set out in regulation 9(3) are satisfied.

242 Air monitoring may also be reduced to once every 12 months where:

(a) RPE is used even though the assessment suggests that it is not strictly required because the OEL is very unlikely to be exceeded;

(b)   it is clearly established that highly effective RPE (in particular, airline breathing apparatus) has to be worn for certain jobs.

243   The circumstances in paragraph 242(b) will occur where provisions for the control of releases of airborne lead cannot, on their own, provide adequate protection, eg jobs involving entry into confined spaces.

244   For the results of a routine air monitoring programme to be effective in protecting the employees' health, it must be possible to compare them with those obtained from previous monitoring exercises. This requires a well-planned programme with, wherever possible, consistency between monitoring exercises to allow results to be compared. This means, for example, that:

(a)   there must be similarity in the process or task monitored;

(b)   monitoring should be carried out at the same stage of the process concerned;

(c)   the same method should be used to collect and analyse the sample(s).

245   If an air monitoring programme is not well planned, it can produce a large number of results without helping to protect employees' health.

Review of air monitoring procedures

246   The air monitoring procedures should be reviewed at intervals in the light of experience, for example, if the accumulated body of data from both air and biological monitoring up to the time of the review suggests that modifications would be justified to the monitoring frequency or to the number of samples obtained for each group or work area for each assessment. Monitoring may highlight deficiencies in control measures and poor work practices. When remedial action has been taken, a further appraisal will be necessary to check whether the changes have had the desired effect. Air monitoring procedures should always be reviewed whenever significant changes are introduced to plant, processes, work methods or dust and fume control provisions which are liable to result in variations in airborne lead concentrations.

Lead alkyls

247   Where lead alkyls are manufactured and handled in concentrated form, established techniques for monitoring using continuous sampling may be used. The standard methods are the US National Institute for Occupational Safety and Health (NIOSH) *Manual of analytical methods,* 4th edition (Department of Health and Human Services (NIOSH) Publication No 94-113), methods 2533 (tetraethyl lead) and 2524 (tetramethyl lead).[15] However, the concentration of lead in the breathing zone of employees should be checked occasionally by using personal sampling.

Monitoring intermittent exposure to lead

248   Monitoring by routine air sampling may not be possible for jobs in which exposures are intermittent, ie where exposures are not on a regular daily or weekly basis. Nevertheless, steps should be taken to make sure that if the exposures are likely to be significant, they are adequately monitored.

Monitoring work at more remote locations

249 Much of the work carried out at remote sites may be irregular and last a relatively short time, and so it may not be possible to carry out routine air monitoring as described above and in HSE's MDHS referred to in paragraph 219.[14] However, if an employee has been assessed as being significantly exposed to lead, air monitoring for a representative sample of the work should be carried out at least every three months.

250 Monitoring may not be necessary for jobs where it is clearly established that RPE has to be worn. However, when these jobs are carried out near to other people who may be affected by the work, air monitoring may be necessary to identify any potential risk to which they may be exposed.

251 It would be prudent to revise the original assessment of the work if the results of measuring the concentrations of lead in employees' blood or urine show that there has been an increased level of lead absorption. In those circumstances, further air monitoring may also be necessary.

Eight-hour time-weighted averages

252 The OELs for lead are assessed over an 8-hour time-weighted average (TWA) reference period. If the total sampling period and the normal working shift is 8 hours then the measured lead-in-air concentrations can be compared directly with the OEL. If the sampling period is less than 8 hours, or the shift is other than 8 hours, the periods over which the samples are taken should be long enough to give results which are representative of normal working exposure. This should include periods of peak exposure, and to allow an 8-hour TWA concentration to be calculated. In most cases this will mean sampling throughout a normal working day or shift and the sampling period should not normally be less than 4 hours.

253 If there is considerable variation in the day-to-day working pattern, it will be necessary to take samples over a number of days to cover the normally expected variations. It is particularly important to take account of activities carried out at the beginning or end of the work period so that activities which may significantly influence the overall exposure level, such as weighing out materials, charging batch mixing machines and machine cleaning, can be included. The results should be reported as 8-hour TWA concentrations. An example of a calculation of 8-hour TWA from full shift samples is set out in Appendix 3.

254 It may be more convenient, and give valuable information, to split the shift into separate shorter duration sampling periods around natural breaks such as meals. The results of these shorter duration samples should be suitably combined on a time-weighted basis to calculate the 8-hour TWA. An example of a calculation of an 8-hour TWA from split samples is also set out in Appendix 3.

255 Further guidance on the design of inhalation exposure monitoring strategies is contained in HSE's *Monitoring strategies for toxic substances*.[16]

Suitable records

256 **Employers are required to keep suitable records of lead-in-air concentrations determined by air monitoring. To be regarded as suitable, a record should state the name and address of the employer and, if appropriate, the site address where the monitoring was carried**

out, and include the following information:

(a) the date of air monitoring;

(b) the name of the employee wearing a personal sampler, the type of work being carried out and, where relevant, its location;

(c) the location of any static sampler;

(d) the length of the sampling time;

(e) analytical results including the 8-hour time-weighted average; and

(f) the names of the sampler and analyst or the names of their organisations.

257 The employer must keep individual exposure and health records for employees where:

(a) personal exposure monitoring is carried out for an identified employee; and

(b) the employee concerned is also under medical surveillance in accordance with regulation 10.

258 The employer may keep the air monitoring records in any format, eg on paper or electronically, but the information should be readily retrievable at any reasonable time and in an easily understood form. It is particularly important that the employer also keeps the information in a form that will help those responsible for medical surveillance to compare the results of air monitoring exposure with any detected effects on the health of employees. Therefore, where an employer keeps an individual employee's personal exposure data and health records on separate electronic databases, the system should be capable of retrieving both sets of employee information so that they can be read and considered together. Alternatively, an employer may keep details of an employee's personal exposure and medical surveillance on the same record.

259 Employers should keep records up to date and retain them for at least five years from the date of the last entry made.

Disposing of records when an employer ceases to trade

260 When an employer or employer's representative, eg an appointed administrator, receiver or liquidator, decides that the business will cease trading, the employer should contact a medical inspector (formerly known as an employment medical adviser) of HSE's Employment Medical Advisory Service (EMAS) at the HSE Area Office nearest to where the business is located, and offer to provide the employees' exposure records (or copies of them) for safe keeping.

Access to employees' records

261 As well as allowing their employees to see their own individual monitoring records, employers may, with the employee's consent, also allow the employee's representatives to see them.

Medical surveillance

(1)   Every employer shall ensure that each of his employees who is or is liable to be exposed to lead is under suitable medical surveillance by a relevant doctor where -

(a)   the exposure of the employee to lead is, or is liable to be, significant;

(b)   the blood-lead concentration or urinary lead concentration of the employee is measured and equals or exceeds the levels detailed in paragraph (2); or

(c)   a relevant doctor certifies that the employee should be under such medical surveillance,

and the technique of investigation is of low risk to the employee.

(2)   The levels referred to in paragraph (1)(b) are -

(a)   a blood-lead concentration of -

(i)   in respect of a woman of reproductive capacity, 20 μg/dl, or

(ii)   in respect of any other employee, 35 μg/dl; or

(b)   a urinary lead concentration of -

(i)   in respect of a woman of reproductive capacity, 20 μg Pb/g creatinine, or

(ii)   in respect of any other employee, 40 μg Pb/g creatinine.

(3)   Medical surveillance required by paragraph (1) shall -

(a)   so far as is reasonably practicable, be commenced before an employee for the first time commences work giving rise to exposure to lead and in any event within 14 working days of such commencement; and

(b)   subsequently be conducted at intervals of not more than 12 months or such shorter intervals as the relevant doctor may require.

(4)   Biological monitoring shall be carried out at intervals not exceeding those set out below -

(a)   in respect of an employee other than a young person or a woman of reproductive capacity, at least every 6 months, but where the results of the measurements for individuals or for groups of workers have shown on the previous two consecutive occasions on which monitoring was carried out a lead in air exposure greater than 0.075 mg/m³ but less than 0.100 mg/m³ and where the blood-lead concentration of any individual employee is less than 30 μg/dl, the frequency of monitoring may be reduced to once a year; or

(b)   in respect of any young person or a woman of reproductive capacity, at such intervals as the relevant doctor shall specify, being not greater than 3 months.

(5)   The employer shall ensure that an adequate health record in respect of each of his employees to whom paragraph (1) applies is made and maintained and

that that record or a copy thereof is kept available in a suitable form for at least 40 years from the date of the last entry made in it.

(6)   *The employer shall –*

(a)   *on reasonable notice being given, allow an employee access to his personal health record;*

(b)   *provide the Executive with copies of such health records as the Executive may require; and*

(c)   *if he ceases to trade, notify the Executive forthwith in writing and make available to the Executive all health records kept by him.*

(7)   *Where the blood-lead concentration for an employee equals or exceeds the appropriate action level, the employer shall take steps to determine the reason or reasons for the high level of lead in blood and shall, so far as is reasonably practicable, give effect to measures designed to reduce the blood-lead concentration of that employee to a level below the appropriate action level.*

(8)   *In any case where the blood-lead concentration or urinary lead concentration of an employee reaches the appropriate suspension level, the employer of that employee shall –*

(a)   *ensure that an entry is made in the health record of the employee by a relevant doctor certifying whether in the professional opinion of the doctor the employee should be suspended from work which is liable to expose that employee to lead;*

(b)   *ensure that a relevant doctor informs the employee accordingly and provides the employee with information and advice regarding further medical surveillance;*

(c)   *review the risk assessment;*

(d)   *review any measure taken to comply with regulation 6, taking into account any advice given by a relevant doctor or by the Executive; and*

(e)   *provide for a review of the health of any other employee who has been similarly exposed, including a medical examination where such an examination is recommended by a relevant doctor or by the Executive.*

(9)   *Further to paragraph (8)(a), where in the opinion of the relevant doctor the employee need not be suspended from work which is liable to expose that employee to lead the entry made in the health record shall include –*

(a)   *the reasons for that opinion; and*

(b)   *the conditions, if any, under which the employee may continue to be employed in such work.*

(10) *Where a relevant doctor has certified by an entry in the health record of an employee that in his professional opinion that employee should not be engaged in work which exposes the employee to lead or that the employee should only be so engaged under conditions specified in the record, the employer shall not permit the employee to be engaged in work which exposes that employee to lead except in accordance with the conditions, if any, specified in the health record, unless that entry has been cancelled by a relevant doctor.*

(11) Where medical surveillance is carried out on the premises of the employer, the employer shall ensure that suitable facilities are made available for the purpose.

(12) An employee to whom this regulation applies shall, when required by his employer and at the cost of the employer, present himself during his working hours for such medical surveillance procedures as may be required for the purposes of paragraph (1) and shall furnish the relevant doctor with such information concerning his health as the relevant doctor may reasonably require.

(13) Where for the purpose of carrying out his functions under these Regulations a relevant doctor requires to inspect any workplace or any record kept for the purposes of these Regulations, the employer shall permit that doctor to do so.

(14) The employer shall ensure that in respect of each female employee whose exposure to lead is or is liable to be significant an entry is made in the health record of that employee by a relevant doctor as to whether or not that employee is of reproductive capacity.

(15) Where an employee or an employer is aggrieved by a decision recorded in the health record by a relevant doctor -

(a) under paragraph (10) that an employee should not be engaged in work which exposes that employee to lead (or which imposes conditions on such work); or

(b) under paragraph (14) that a female employee is of reproductive capacity,

the employee or employer may, by an application in writing to the Executive within 28 days of the date upon which the decision was notified to the employee or employer as the case may be, apply for that decision to be reviewed in accordance with a procedure approved by the Health and Safety Commission, and the result of that review shall be notified to the employee and employer and entered in the health record in accordance with the approved procedure.

Suitable medical surveillance

262 Where exposure to lead is significant as defined in regulation 2, the employer should make sure that the employee is under medical surveillance by either a medical inspector (formerly known as an employment medical adviser) of HSE's Employment Medical Advisory Service (EMAS) or appointed doctor, ie the 'relevant doctor'. Suitable surveillance procedures comprise initial and periodic medical assessments which include measuring the employee's blood-lead and/or urinary lead concentration. Suitable measurement methods are listed in Appendix 4 but the method prescribed by the Regulations for measuring the concentration of lead in blood or urine is atomic absorption spectroscopy (see the definition of 'biological monitoring' in regulation 2).

263 Regulation 10(1)(c) requires that an employee who is not significantly exposed to lead should nevertheless be placed under medical surveillance if the relevant doctor certifies it as necessary. For example, the doctor may consider this to be a sensible and cautious step to take if observation of an employee's poor work practices or poor standards of personal hygiene suggests that there is a greater risk of that employee absorbing lead.

264   The objectives of medical surveillance are to:

(a)   make an initial assessment of an employee's suitability to carry out work with lead;

(b)   evaluate the effect of lead absorbed by employees and to advise them on their state of health;

(c)   monitor the exposure of female employees of reproductive capacity;

(d)   assess the suitability of an employee to carry on working where there is continuing exposure to lead;

(e)   detect early signs of excessive lead absorption or early adverse health effects, and to remove employees from exposure to prevent lead poisoning and other health effects developing; and

(f)   help employers in their duty to control the exposure of their employees to lead.

265   Employers should advise women capable of having children on the special need to protect any developing foetus. When a woman declares that she is pregnant, the employer should take appropriate action to remove her from work where her exposure to lead is significant and comply with any other requirements of the Management of Health and Safety at Work Regulations 1999.

266   An initial medical assessment should always be carried out on all new employees employed in an activity liable to expose them to lead, and who have been exposed to lead at work in a previous job in the last three months, irrespective of whether their exposure to lead in their new employment is likely to be significant. Where the employee's blood or urine reveals a level of lead equivalent to or greater than the appropriate concentration set out in regulation 10(2), the employee concerned should be regarded as having been significantly exposed to lead and be placed under medical surveillance. The subsequent monitoring of the employee's blood-lead or urinary lead concentrations should be carried out in accordance with the guidance in paragraphs 276-278. Monitoring should continue until the employee's blood-lead or urinary lead drops below the appropriate concentration set out in regulation 10(2) or until the relevant doctor certifies that it is no longer necessary.

Woman of reproductive capacity

267   For all female employees, the employer must decide whether the employee is of reproductive capacity. This is necessary to determine which action and suspension levels should be applied to her.

268   A woman of reproductive capacity is one who is medically and physically capable of becoming pregnant. A woman's marital or other status, whether her husband or partner is sterile etc are not relevant factors in determining this. The all-important consideration that should be applied is whether a woman has a medical or physical condition that would make it impossible for her to conceive, eg she is sterilised, or had a hysterectomy, or is clearly post-menopausal.

269   For practical purposes, an employer may find it easier to regard a female employee as being of reproductive capacity unless there is clear

and obvious evidence to show otherwise. If for any reason an employer cannot easily determine whether a woman is capable of becoming pregnant, this decision should be made in consultation with the relevant doctor. For a female employee whose exposure to lead is likely to be significant, the decision on whether she is of reproductive capacity should be recorded in her health record in accordance with regulation 10(14). The decision should be reviewed at appropriate intervals and in the light of changed circumstances. (See also Appendix 5 'Guidance notes for appointed doctors on the Control of Lead at Work Regulations 2002'.)

Lead and lead compounds except lead alkyls

Initial medical assessment

270   The employer should notify the doctor of the name of each person newly employed or whom it is intended to employ on work which is likely to expose that person significantly to lead so that the doctor can carry out the initial medical assessment.

271   The initial medical assessment should be carried out so far as is reasonably practicable before a person starts work for the first time which is likely to result in significant exposure to lead, and in any event not later than 14 working days after first exposure. The assessment should consist of:

(a)   consideration of the employee's occupational record with particular reference to any earlier exposures to lead, and any previous suspensions;

(b)   a clinical assessment including consideration of medical history, clinical conditions, and personal hygiene and intellectual capacity to work with hazardous substances;

(c)   measurement of 'baseline' blood-lead and haemoglobin.

272   The doctor may also want to verify the results of initial baseline measurements by carrying out some of the further biological tests specified in Appendix 4.

Periodic medical assessments

273   These should be carried out and consist of:

(a)   measurement of blood-lead concentrations;

(b)   at least once a year, a clinical assessment to include:

(i)   a review of the employee's medical records and occupational history;

(ii)   a physical examination giving special attention to the symptoms associated with the ill-health effects of lead exposure and of early lead poisoning;

(iii) consideration of whether there are any observable trends or patterns in the employee's blood-lead level, and how these might relate to work practices, personal hygiene, changes in exposure and any current ill health or recent sickness absence.

274 At the doctor's discretion, other relevant biological tests may be carried out annually such as measuring the employee's haemoglobin and/or zinc protoporphyrin (ZPP) (see Appendices 4 and 5).

275 The intervals between periodic medical assessments should not exceed 12 months. The relevant doctor may decide the frequency of carrying out periodic medical assessments as long as the employee's blood-lead concentration remains below the appropriate suspension level.

Monitoring an employee's blood-lead concentration

276 When employees are significantly exposed to metallic lead and its compounds, their blood-lead levels should be measured every three months. If exposure is uniform, then a consistent blood-lead pattern will probably be established, although this may take several months. Thereafter, the interval can be extended according to Table 3, except for women of reproductive capacity and for young persons for whom the interval should not be extended to more than three months.

Table 3 Intervals for blood-lead measurements for various categories

Category	Blood-lead µg/dl	Maximum interval between blood-lead measurements
A	under 30	12 months (see note below)
B	≥ 30 and < 40	6 months
C	≥ 40 and < 50	3 months
D	≥ 50 and < 60	3 months
E	60 and over	At the doctor's discretion but not more than three months

Category A indicates that the absorption of lead due to occupational exposure is reasonably well controlled. The interval between blood-lead measurements should not, however, be longer than six months unless on the previous two consecutive occasions on which air monitoring was carried out, measurement of the airborne lead to which that person was exposed was less than 0.10 mg/m^3.

Category B indicates that lead is being absorbed due to occupational exposure to lead. For employees in this category other suitable biological tests may also be carried out in addition to six-monthly blood-lead measurement. Suitable biological tests include measurement of zinc protoporphyrin (ZPP), erythrocyte protoporphyrins, aminolaevulinic acid dehydratase in blood (ALAD) and aminolaevulinic acid in urine (ALAU) and may be carried out every 12 months (see also Appendix 4).

Category C also indicates that lead is being absorbed due to occupational exposure but at a higher concentration than for employees in Category B, and that blood-lead concentrations may be approaching the action level. Other suitable biological tests may also be appropriate as for Category B employees.

Category D indicates that the blood-lead concentrations have breached the action level and the employer should carry out an investigation in accordance with paragraphs 281 and 282. This range of blood-lead concentrations also represents the level at which the employee should come under direct medical surveillance in that a clinical assessment and any other relevant biological tests should be carried out as soon as possible after the blood-lead concentration has been confirmed. The clinical examination may be deferred until measurement of the blood-lead concentration carried out at a time determined by the doctor shows that the action level of 50 µg/dl continues to be breached.

Category E represents the concentration at or above which the doctor may certify the employee as unfit for work where there is liable to be exposure to lead.

277 Some types of exposures such as the burning of lead paint or lead-covered metal during demolition work, scrap metal work, blast removal of old lead paint, notably on railway ironwork, road bridges and similar structures etc are likely to be so variable that a clear pattern of lead absorption cannot be established. In these situations it may be necessary to continue three-monthly blood-lead tests for as long as medical surveillance is required, or even more frequently if the doctor considers it necessary.

278 The results of the initial blood-lead measurement made before or within 14 working days of an employee starting work should be used as a baseline for comparison with the first follow-up measurement taken after a suitable period of work involving significant exposure to lead. This first follow-up blood-lead measurement should be made no more than three months after the initial one. However, where employees are carrying out work involving variable exposure to lead such as described in paragraph 277, the doctor may decide that the time between the initial and first follow-up measurement should be considerably reduced, eg to two weeks. The doctor should then decide the frequency of further tests having regard to the results obtained.

Action levels

279 All employees who are likely to be significantly exposed to lead at work, other than lead alkyls, should be made subject to a blood-lead concentration action level. The purpose of the action level is to:

(a) warn the employer that an employee's blood-lead concentration is approaching the suspension level;

(b) prompt the employer to investigate why it has been breached and to review the range and effectiveness of control measures used with the aim of reducing the employee's blood-lead below the action level; and

(c) prevent the employee's blood-lead concentration from reaching the suspension level.

280 The action levels for different groups of employees are set out in Table 4.

Table 4 Action levels for different groups of employees

Group of employees	Blood-lead concentration µg/dl
Women of reproductive capacity	25
Young persons (aged 16 and 17)	40
Any other employee	50

281 In investigating why the action level has been breached, the employer's review should include the following:

(a) a check that recommended and established work practices are being followed;

(b)   a check on the effectiveness of all control measures, including where appropriate that engineering controls are working as they should to their design specification and do not need repair;

(c)   where appropriate, any RPE is being properly used in accordance with instructions;

(d)   the employee(s) is following strict hygiene procedures; and

(e)   consultation with the doctor to agree any additional protective or preventive measures to be taken for any selected individuals or groups of employees.

282   If recent air monitoring results are not available, the employer should also consider whether there is a need to measure the concentration of lead-in-air in the breathing zones of the employees concerned.

283   An employee's blood-lead level may remain above the action level despite steps taken in response to the investigation carried out under paragraph 281. If so, the employer should keep a record of any remedial action taken and repeat the investigation within three months of the first one.

Suspension levels

284   All employees who are liable to be exposed to lead at work are subject to a suspension level. This is the blood-lead concentration (or urinary lead concentration where there is exposure to lead alkyls) at which the doctor decides whether to certify that the employee should no longer be exposed to lead. The suspension levels for different groups of employees exposed to metallic lead and its compounds are set out in Table 5.

Table 5 Suspension levels for different groups of employees

Group of employees	Blood-lead concentration µg/dl
Women of reproductive capacity	30
Young persons (aged 16 and 17)	50
Any other employee	60

285   Any employee whose blood-lead concentration reaches the appropriate suspension level should have the test repeated *urgently*. Employers should ensure that every effort is made to obtain the result of the repeat test within *ten working days* of the initial result becoming available. Where the initial result relates to a young person or a woman of reproductive capacity, or it reaches or exceeds 70 µg/dl for all other employees, the employer should consult the relevant doctor and consider whether to remove the employee concerned from work involving exposure to lead until the result of the repeat test becomes available. If the result of the repeat test is also equal to or greater than the appropriate suspension level, the doctor should certify that the employee be taken off work which further exposes the employee to lead.

286   However, some employees, excluding women of reproductive capacity, who have worked for many years in the lead industry, may have built up a high body burden of lead which could take a long time to

fall below their suspension level of 60 µg/dl. Specifically, these employees are those who either:

(a)   have been employed on work which exposed the employee to lead for at least 20 years; or

(b)   are aged 40 years or more and may have been employed on work involving exposure to lead for at least 10 years.

287   As a special concession to these longer-term employees, the employer may take some additional factors into account in deciding whether they should be taken off work involving exposure to lead.

(a)   Employees who satisfy either of the conditions (a) or (b) of paragraph 286 *before* the Control of Lead at Work Regulations 2002 came into force may continue in work which exposes them to lead as long as their blood-lead concentration is not more than 80 µg/dl and the ZPP level remains lower than 20 µg/g haemoglobin, or the ALAD level remains greater than 6 European units or the ALAU level remains lower than 20 mg/g creatinine.

(b)   Employees who satisfy either of the conditions (a) or (b) of paragraph 286 *on or after* the Control of Lead at Work Regulations 2002 came into force may continue in work which exposes them to lead as long as their blood-lead concentration is not more than 70 µg/dl and the ZPP level remains lower than 20 µg/g haemoglobin, or the ALAD level remains greater than 6 European units or the ALAU level remains lower than 20 mg/g creatinine.

288   For these longer-term employees who continue in work under these circumstances, the employer should nevertheless make every effort to reduce the employee's blood-lead concentration to below 60 µg/dl, and the doctor should consider increasing the frequency of blood-lead and haemoglobin testing.

289   The doctor may use discretion to certify whether an employee should be taken off work which further exposes that employee to lead even where the employee's blood-lead concentration is below the suspension level, for example:

(a)   where an employee shows symptoms of anaemia which might not be related to lead exposure but needs to be investigated to establish the cause; or

(b)   where the results of other biological tests or a clinical assessment suggests such action is necessary pending further investigation.

Women of reproductive capacity

290   If the blood-lead concentration of a woman of reproductive capacity triggers the suspension level but the doctor does not certify her as unfit to continue in work which exposes her to lead, the doctor should record the reasons in the employee's health record, eg the woman has left the employment concerned.

291   To safeguard any developing foetus, if a woman of reproductive capacity becomes pregnant, she should notify her employer as soon as possible in accordance with the requirements of the Management of Health and Safety at Work Regulations 1999. The employer should in turn notify the doctor of the pregnancy and, on the advice of the doctor, remove the pregnant employee from any work involving significant exposure to lead.

Lead alkyls

292   The method for monitoring the absorption of lead alkyls in the body is by measuring total urinary lead concentration, and the concentration of analytes in urine is corrected for the urinary creatinine concentration to allow for differences in the volume of urine produced. The urine analyte concentration is then corrected to what it would be if there were 1 g/litre of creatinine in the urine, and values are then expressed in units of µg Pb/g creatinine. This biological monitoring method should be used for carrying out initial and periodic medical assessments on employees exposed to these particular lead compounds.

293   The general arrangements described in paragraphs 270-275 for carrying out initial and periodic medical assessments also apply to employees who are exposed to or who work with lead alkyls. However, at least once a year the medical assessment should consist of the measurement of blood-lead as well as of urinary lead concentrations and, where the results of these indicate the need for it, a clinical assessment.

294   The intervals between carrying out periodic urinary lead measurements should be determined as set out in Table 6.

Table 6 Intervals for various urinary lead measurements

Group of employees	Urinary lead as µg Pb/g creatinine	Maximum interval between urinary lead measurements
All employees including young persons (aged 16 and 17) except women of reproductive capacity	Under 95 95*-109 110 and over**	6 weeks 1 week At the doctor's discretion
Women of reproductive capacity	under 20 20*-24 25 and over**	6 weeks 1 week At the doctor's discretion

Note: *   Represents the level at which the employee will come under closer medical surveillance.
        **   Represents the level above which the doctor will certify the employee as unfit for work which exposes the employee to lead.

Action levels

295   There are no action levels for employees exposed to lead alkyls.

Suspension levels

296 The suspension levels for employees exposed to lead alkyls are set out in Table 7.

Table 7 Suspension levels

Group of employees	Urinary lead concentration: μg/dl Pb/g creatinine
Women of reproductive capacity	25
Young persons (aged 16 and 17)	110
Any other employee	110

297 Any employee, other than a woman of reproductive capacity, whose urinary lead concentration is equal to or greater than 110 μg Pb/g creatinine, or a woman of reproductive capacity whose urinary lead concentration is equal or greater than 25 μg Pb/g creatinine, should have the test repeated *urgently*. Employers should ensure that every effort is made to obtain the result of the repeat test *within ten working days* of the initial result becoming available. Where the initial result relates to a woman of reproductive capacity, the employer should consult the relevant doctor to consider whether to remove the employee concerned from work involving exposure to lead alkyls until the result of the repeat test becomes available. If the result of the repeat test confirms that the suspension level has been reached or exceeded, the doctor should certify that the employee be taken off work which exposes the employee to lead alkyls.

298 However, the doctor may use discretion to certify whether an employee should be removed from work involving further exposure to lead alkyls even where the employee's urinary lead concentration is below the suspension level, eg where the results of other biological monitoring tests and/or a clinical assessment that the doctor may have carried out indicates a need to suspend the employee pending further investigations.

299 Where jobs likely to result in exposure to lead alkyls are carried out infrequently, eg inspections by managerial or technical employees or cleaning of leaded petrol tanks by contractors' employees, the doctor should decide the frequency of carrying out medical assessments.

Action when an employee is certified as unsuitable for work with lead

300 When the doctor certifies that an employee should be removed from any work which exposes that employee to lead, it is the responsibility of the employer to make sure that there is compliance with the certificate. The employer should make every effort to preserve the employee's terms and conditions of employment and to redeploy the employee concerned on suitable alternative work that complies with any conditions certified by the doctor in the employee's health record. Where the employer cannot provide alternative employment and has to suspend the employee from work, the employee should not return to work which would involve exposure to lead until the doctor considers it appropriate for the employee to do so.

301 The decision on when an employee can return to work involving exposure to lead should only be authorised by the doctor. This should be a matter of the doctor's judgement, taking account of:

(a) the employee's degree and duration of exposure to lead;

(b) recent blood-lead or urinary lead levels;

(c) the results of any other biochemical tests;

(d) the prevailing conditions in the place of work; and

(e) for a woman who returns to work after giving birth, whether or not she is still breastfeeding her child.

302 At the doctor's discretion, the employee may return to work where exposure to lead is not significant, eg in another section of the factory. Before approving this, the doctor should be aware of the conditions prevailing in the place of work concerned, eg the expected lead-in-air levels and the general cleanliness of the work area.

Liaising with a suspended employee's own doctor

303 When the amount of lead in an employee's blood or urine results in their temporary removal from work involving exposure to lead, the relevant doctor should seek the employee's consent and write to tell the employee's own doctor (general practitioner) so that the GP can make a note in the employee's medical records.

Providing information

304 The results of medical surveillance not only provide a means of monitoring lead absorption in employees, but also reflect the effectiveness of the hygiene provisions and control measures and the employee's standard of personal hygiene. Therefore, it is important that the results of medical surveillance are made available to the employer in a form that identifies areas where:

(a) control measures are not working effectively; or

(b) improvements or extra measures are needed, or supervision needs to be strengthened.

305 Therefore the employer should make sure that as a part of normal medical surveillance, the doctor:

(a) tells the employee under surveillance of the results and the significance of his/her clinical assessment; the results of measuring blood- or urinary lead concentration and any other biological tests as soon as possible when the results are unsatisfactory, and in other cases where the employee examined requests the information;

(b) tells the employer immediately of the names of employees whose blood-lead concentrations have reached or exceed the appropriate action level so that the employer can pursue action as set out in paragraph 281;

(c) tells the employer immediately of the names of employees who should be removed from further exposure to lead or who may continue working only under specified conditions. Where medical surveillance is carried out by an appointed doctor (as distinguished from a medical inspector), the doctor should forward a copy of the appropriate record (ie form FOD MS 103: 'Certification of unfitness/fitness' - see Annex 4 of Appendix 5) to the medical inspector at the HSE Area Office nearest to the location of the business within seven days of certifying any employee as unfit so that the enforcing authorities can be alerted. The appointed doctor should also advise the medical inspector of any unsatisfactory trends or matters of concern;

(d) makes available to the employer, in writing, the results of measuring blood/urinary lead concentrations and of any other biological tests for all employees who have given their consent for the information to be divulged. However, the employer should always be advised of the results in enough detail for them to assess the levels of absorption so that adverse trends in groups, individuals or specific work areas can be remedied or adverse conditions rectified;

(e) provides the employer with a record of all employees who the doctor certifies should be suspended from work with lead, or for whom conditions have been specified about their continuing employment in work exposing them to lead. The record should show:

  (i)   the employee's name;

  (ii)   the employee's job;

  (iii) the date of the medical assessment;

  (iv)   certification that the employee should be removed from work involving any exposure to lead;

  (v)   conditions specified for the employee's return to work with lead;

  (vi)   the dated signature of the doctor; and, where appropriate,

  (vii) authorisation by the doctor of the employee's fitness to return to work with lead.

306 At the doctor's discretion, the notifying of medical information to the employer may be delegated to a nurse or other occupational health specialist who works closely under the doctor's supervision.

Suitable facilities for carrying out medical surveillance

307 Where medical surveillance procedures, eg medical examinations and blood or urine sampling, are carried out at the employer's premises, suitable facilities should be made available. These should comprise a room which is:

(a) properly cleaned, adequately warmed and well ventilated;

(b) suitably furnished with a table and seats;

(c) equipped with a washbasin equipped with hot and cold running water, soap and a clean towel. If it is not reasonably practicable to provide hot and cold running water, then a supply of warm water should be provided.

308 The room should be set aside for the exclusive purpose of medical surveillance whenever it is required, and it should provide privacy. Where the number of employees to be examined or assessed is substantial, then where reasonably practicable, a suitable waiting area should be provided. An adjacent or nearby toilet with handwashing facilities should be available for employees.

Suitable records

309 The employer should make sure that basic details are recorded of all people who need to be under medical surveillance. The information the record should contain is summarised in Appendix 6.

310 Once a year the appointed doctor should submit to HSE a statistical return on the number of people under medical surveillance with information covering: their gender; the lead sector in which they are employed; the distribution of their blood-lead measurements; and the number and the reasons why they are suspended from lead work. The statistical return should identify separately young men and women aged under 18.

311 Medical surveillance records should be kept available for at least 40 years as they could provide useful occupational hygiene data.

Implications of the Data Protection Act

312 The Data Protection Act 1999 places requirements on employers and others who hold information on medical surveillance records. In particular, those holding such records must take steps to tell those on whom records are held:

(a) that a record is being kept;

(b) the purpose for which it is being held; and

(c) that they have the right of access to the information and the right to correct it.

313 Under data protection law, the people on whom information is held must be given access to their records and the right to have any inaccurate information corrected. The contents of individual records can only be disclosed in line with the purposes for which the information is kept. It is lawful to disclose information to HSE about employees placed under medical surveillance under the Control of Lead at Work Regulations 2002 because of their significant exposure to lead.

Disposing of records when an employer ceases to trade

314 When an employer or employer's representative, eg an appointed administrator, receiver or liquidator, decides that the business will cease trading, the employer should contact a medical inspector at the HSE

Area Office nearest to where the business is located, and offer to provide the employees' exposure records (or copies of them) for safe keeping.

Access to employees' records

315   As well as allowing their employees to see their own individual blood-lead or urinary lead monitoring records maintained under regulation 10(5), employers may, with the employee's consent, also allow the employee's representatives to see them. Where HSE makes a request under regulation 10(6)(b) for copies of employees' personal health records, the employer should only provide the information summarised under Appendix 6.

Applying for a review of a decision recorded in the employee's health record

316   Regulation 10(15) gives both an employee and employer a statutory right to apply to HSE for a review of a decision recorded in the employee's health record on the points covered in sub-paragraphs (a) and (b) of the regulation. Applications should be sent to:

Head of Division
Health Directorate C
Health and Safety Executive (Level 7, North Wing)
Rose Court
2 Southwark Bridge
London SE1 9HS

317   Applications should include the following basic information:

(a)   the full name of the applicant;

(b)   their address;

(c)   their date of birth;

(d)   their gender;

(e)   their place of employment and their job;

(f)   the date on which the medical examination in question took place;

(g)   if possible, a summary obtained from the appointed doctor as to the reasons why the doctor came to the decision that is the subject of the appeal;

(h)   the applicant's reasons why the doctor's decision is considered to be wrong and why it should be reviewed.

318   An employee whose removal from exposure to lead becomes the subject of an application for review to HSE should not carry out any further work liable to expose that employee to lead until the outcome of the review has been decided.

Medical surveillance at more remote locations

319   When employees work with or are exposed to lead at more remote

locations, medical surveillance should follow the general procedure described in paragraphs 262-321. However, there will be some work activities when an employee could be exposed to high concentrations of lead dust or fume with the risk of very rapidly absorbing lead, eg welding or cutting lead-painted structures.

320 In these circumstances, the frequency of measuring blood-lead concentrations may need to be increased or the samples taken to reflect the exposure to lead. The relevant doctor should decide when clinical assessments and measuring blood-lead concentrations should be carried out in the light of the information provided by the employer.

321 It is important, therefore, that the employer gives the doctor enough information about the work programme and possible exposure levels to allow the doctor to make these decisions.

Regulation 11

Information, instruction and training

(1) *Every employer who undertakes work which is liable to expose an employee to lead shall provide that employee with suitable and sufficient information, instruction and training.*

(2) *Without prejudice to the generality of paragraph (1), the information, instruction and training provided under that paragraph shall include -*

(a) *details of the form of lead to which the employee is liable to be exposed including -*

(i) *the risk which it presents to health,*

(ii) *any relevant occupational exposure limit, action level and suspension level,*

(iii) *access to any relevant safety data sheet, and*

(iv) *other legislative provisions which concern the hazardous properties of that form of lead;*

(b) *the significant findings of the risk assessment;*

(c) *the appropriate precautions and actions to be taken by the employee in order to safeguard himself and other employees at the workplace;*

(d) *the results of any monitoring of exposure to lead carried out in accordance with regulation 9; and*

(e) *the collective results of any medical surveillance undertaken in accordance with regulation 10 in a form calculated to prevent those results from being identified as relating to a particular person.*

(3) *The information, instruction and training required by paragraph (1) shall be -*

(a) *adapted to take account of significant changes in the type of work carried out or methods of work used by the employer; and*

(b) *provided in a manner appropriate to the level, type and duration of exposure identified by the risk assessment.*

(4)   Every employer shall ensure that any person (whether or not his employee) who carries out work in connection with the employer's duties under these Regulations has suitable and sufficient information, instruction and training.

(5)   Where containers and pipes for lead used at work are not marked in accordance with any relevant legislation listed in Schedule 2, the employer shall, without prejudice to any derogations provided for in that legislation, ensure that the contents of those containers and pipes, together with the nature of those contents and any associated hazards, are clearly identifiable.

Suitable and sufficient information, instruction and training

322   The information, instruction and training given to employees, including supervisors, should:

(a)   take account of the findings of the assessment and in particular the degree of risk it reveals;

(b)   be provided in whatever form is most suitable for the intended audience, ie it may range from quite simple oral communication with a group of employees, to more comprehensive instruction and training supported by written information given to individual employees; and

(c)   be understandable.

323   For the information, instruction and training provided to be suitable and sufficient, it should cover in appropriate detail the topics set out in regulation 11(2) and also include the following:

(a)   the health hazards of lead, including the health hazards to employees' families and others which could result from taking home contaminated clothing and equipment;

(b)   how control measures such as engineering controls, work methods, personal protective equipment and their correct use can reduce the risks by limiting the employee's exposure to lead and the spread of lead dust and fume to as low a level as is reasonably practicable, and the reasons for using these in relation to the employee's own job;

(c)   the use and maintenance of control measures; the use of protective clothing and, where appropriate, respiratory protective equipment in accordance with any recommendations and instructions supplied by the manufacturer; and the procedures for reporting and correcting defects;

(d)   the assessment of exposure, the purpose of air monitoring and the meaning of significant exposure and its relationship with other measures in the Regulations;

(e)   the role of medical surveillance and especially of biological monitoring and the action and suspension levels; the duties of employees to attend for medical examinations and biological tests at the appointed date and time;

(f) when to use the hygiene facilities provided and the importance of doing so in accordance with agreed procedures, and the need for employees to practise a high standard of personal hygiene;

(g) the procedures for dealing with accidents, incidents or emergencies involving lead, eg the instructions set out in the employer's procedures to comply with regulation 12 (where appropriate) to deal with an uncontrolled release of lead dust or fume into the workplace. Employers should ensure that all employees have the opportunity to read and discuss the procedures with their safety representatives; and

(h) any further relevant information resulting from a review of the assessment: why it has been done and how any changes will affect the way employees do the work in the future.

324 Regulation 11(2)(a)(iii) allows employees, including safety representatives who are employees, access to any relevant safety data sheet. These sheets are not always the best or most appropriate way of providing employees with information about the risks associated with the use of a substance or compound containing lead in the workplace. Employers may instead choose to distil the information from safety data sheets onto more readable and understandable in-house information and training documents. In doing so, employers should take every care in transcribing the information and in amplifying warnings and precautions accurately. While this is an acceptable practice, employees and their safety representatives must still be allowed access to safety data sheets relevant to the work should they want to see them.

325 The other legislative provisions referred to in regulation 11(2)(a)(iv) should apply directly to a lead compound. These may include the Chemicals (Hazard Information and Packaging for Supply) Regulations 2002 (CHIP).

326 The employer should issue a copy of HSE's free leaflet *Lead and you*[17] to all employees on their first employment on work with lead, and make copies available for issue at the request of any employees or their representative.

327 Employers need to take particular care to make sure that adequate information, instruction and training is given to employees who:

(a) work either alone or in small numbers at more remote locations; and

(b) are relied upon to supervise their own work.

328 The extent of the information, instruction and training will vary with the complexity of the hazards, risks, processes and controls which the risk assessment will identify. Employers should aim to strike a balance between providing sufficient information for an employee to carry out work safely, and providing too much information that may result in overburdening and confusing the employee. Basic instructions and training may be all that is required for some activities where exposure to lead is unlikely to be 'significant'.

329 Employers have a duty under the Management of Health and Safety at Work Regulations 1999 to ensure that the information they provide is

comprehensible. Employers should consider all the various ways of providing information, instruction and training and select those most appropriate to their own circumstances. The range of options includes: class or group tuition, individual tuition, written instructions including leaflets, courses etc. Employers must also decide how much time is needed to provide suitable and sufficient training etc for their employees to comply fully with the detailed requirements of the Management Regulations. New employees must be provided with proper induction training which should always cover emergency and evacuation procedures.

Updating information etc

330 Providing information, instruction and training is not a one-off exercise. Information, instruction and training should be reviewed and updated whenever significant changes are made to the type of work carried out or to the work methods used. Significant changes might include the amount of lead used or produced, new control measures, new compounds of lead brought into the workplace, and/or automation of certain processes. Further information and training following a review of the assessment should cover why the assessment was reviewed, any changes to the way the work is to be done and the precautions the employees should take to protect themselves and others.

Making information available to safety representatives

331 The employer must make all information available to employees or their representatives in accordance with the Health and Safety (Consultation with Employees) Regulations 1996, and the Safety Representatives and Safety Committees Regulations 1977.

Instruction and training

332 The instruction and training must ensure that people at work on the premises do not put themselves or others at risk through exposure to lead. In particular, the instruction must be sufficient and suitable for them to know:

(a) how and when to use the control measures;

(b) the defined methods of work;

(c) how to use the PPE and especially RPE, eg the correct method of removing and refitting masks, gloves etc;

(d) the cleaning, storage and disposal procedures they should follow, why they are required and when they are to be carried out, eg cleaning contaminated PPE with water or a vacuum fitted with a HEPA filter, and not with an airline, and the risks of using contaminated PPE;

(e) the procedures to be followed in an emergency.

333 Training should include elements of theory as well as practice. Training in the use and application of control measures and PPE should take account of any recommendations and instructions supplied by the manufacturer.

Records of training

334 Employers may find it helpful to keep a record of the training given to individual employees or specific groups of employees. The records may provide a useful checklist for ensuring that employees receive all the necessary training at the appropriate time. The records may also help to resolve any disputes that may arise about whether the employer has provided a particular employee with a specific aspect of information, instruction and training.

Providing employees with written records

335 As a matter of good practice, the employer should provide employees with a written record of the following information when their employment ends for any reason:

(a) the concentrations of lead-in-air to which they have been exposed with appropriate dates;

(b) the results of their own lead in blood or urine measurements; and

(c) an explanation of the significance of these results.

336 The employer should advise the employee to keep the records indefinitely in a secure place.

People who carry out work on behalf of the employer

337 Anyone who carries out any work on behalf of the employer in relation to their duties under regulation 11(4) should possess the knowledge, skill and experience to be able to perform that work effectively. The employer should therefore make sure that the person to whom work is delegated is competent to do it. This applies particularly to anyone who carries out an assessment, maintains control measures or conducts air monitoring which requires the use of instruments or special equipment. In these circumstances, the person concerned should be given additional information, instruction and training as appropriate on:

(a) the purpose of using instruments and/or the special equipment;

(b) the correct use of instruments and/or the special equipment;

(c) sampling techniques; and

(d) the analysis, interpretation and recording of results.

338 Wherever practicable, employers should encourage suitable employees to have the training, and to gain the knowledge and experience that will give them the competence to help their employers comply with the requirements of the Regulations. If it is necessary to use expertise from outside the business, the employer will still need to make sure that the people engaged receive sufficient information about the particular circumstances of the work. HSE does not define or approve standards of competence, but employers seeking expert advice may wish to consult a professional occupational hygienist. The British Institute of Occupational Hygienists (BIOH) is recognised by HSE for its occupational hygiene qualifications and it maintains a register of competent practitioners. For the address and telephone number of BIOH, see paragraph 53.

339 The methods the employer uses to achieve all the objectives covered by regulation 11 should preferably be combined with and be a normal part of any information, instruction and training which is given to employees and others so that they can carry out their job, eg as part of practical training on how to use particular plant or equipment. In some cases formal training by means of seminars, lectures, discussion groups or written and oral presentation as well as by demonstration could be appropriate, or on-the-job training by an experienced and knowledgeable supervisor. Employers should also arrange for refresher training to be given where it is necessary to ensure that essential information and skills etc are not forgotten, and the overall objectives of having an informed and trained workforce are met.

Identifying the contents of containers and pipes

340 Regulation 11(5) is intended to cover those circumstances where there is no other legal requirement for employers to identify containers and pipes which contain lead or lead compounds etc. These legal requirements are listed in Schedule 2 which implement European Union Directives containing requirements for the hazardous contents of containers and pipes to be clearly identified and, in some circumstances, labelled.

341 Many containers will already be adequately marked with their contents because of supply requirements. Similarly, pipework is often marked in accordance with BS 1710: 1984 *Specification for identification of pipelines and services*[18] or equivalent 'in-house' standards. However, there are containers and pipes whose contents are not individually marked because it is not practicable to do so, eg the pipework at large chemical complexes whose contents may change frequently during the course of a process or operation.

342 Employers should ensure that they have suitable procedures in place to identify all containers and pipes containing lead or lead compounds etc. The identification process may involve reference to working procedures, to operating instructions etc which identify individual plant components by name or number.

343 Complex plant or batch processing may require components to be used for lead or lead compounds as well as other hazardous substances over short periods of time, and in these circumstances, employees will need to be familiar with the plant operations and the sources of information available to them. Whichever identification procedure employers adopt, they must ensure that employees and safety representatives are familiar with any plans, characters, signs, symbols, codes etc that the identification system or procedures use.

344 For some repair or maintenance work involving the opening of vessels or breaking into pipework used for lead or lead compounds, it may be necessary for the work to be carried out under the control of a permit-to-work system. For this work, identifying the containers or pipes which contain or contained lead is one essential element of the risk assessment that must be carried out before the work starts. For employers to carry out suitable and sufficient assessments of the work in these circumstances, they must have a system in place which identifies:

(a) the form the lead takes, eg liquid, semi-liquid, sludge, powder, waste mixed with other identified material;

(b) the hazards the lead could pose if employees were exposed to it.

Arrangements to deal with accidents, incidents and emergencies

(1) Subject to paragraph (4) and without prejudice to the relevant provisions of the Management of Health and Safety at Work Regulations 1999[(a)], in order to protect the health of his employees from an accident, incident or emergency related to the presence of lead at the workplace, the employer shall ensure that -

(a) procedures, including the provision of appropriate first-aid facilities and relevant safety drills (which shall be tested at regular intervals), have been prepared which can be put into effect when such an event occurs;

(b) information on emergency arrangements, including -

(i) details of relevant work hazards and hazard identification arrangements, and

(ii) specific hazards likely to arise at the time of an accident, incident or emergency,

is available; and

(c) suitable warning and other communication systems are established to enable an appropriate response, including remedial actions and rescue operations, to be made immediately when such an event occurs.

(2) The employer shall ensure that information on the procedures and systems required by paragraph (1)(a) and (c) and the information required by paragraph (1)(b) is -

(a) made available to relevant accident and emergency services to enable those services, whether internal or external to the workplace, to prepare their own response procedures and precautionary measures; and

(b) displayed at the workplace, if this is appropriate.

(3) Subject to paragraph (4), in the event of an accident, incident or emergency related to the presence of lead at the workplace, the employer shall ensure that -

(a) immediate steps are taken to -

(i) mitigate the effects of the event,

(ii) restore the situation to normal, and

(iii) inform those of his employees who may be affected; and

(b) only those persons who are essential for the carrying out of repairs and other necessary work are permitted in the affected area and they are provided with -

(i) appropriate personal protective equipment, and

(ii) any necessary specialised safety equipment and plant,

which shall be used until the situation is restored to normal.

(a) SI 1999/3242.

(4)   Paragraphs (1) and (3) shall not apply where -

(a)   the results of the assessment show that, because of the quantity of lead present at the workplace, there is only a slight risk to the health of employees; and

(b)   the measures taken by the employer to comply with the duty under regulation 6(1) are sufficient to control that risk.

General

345   The requirements in regulation 12 of CLAW are in addition to those contained in regulation 8 of the Management of Health and Safety at Work Regulations 1999 (the Management Regulations). The Management Regulations impose a number of general duties on all employers to establish procedures to deal with situations involving serious and imminent danger. Details of these are described in *Management of Health and Safety at Work Regulations 1999. Approved Code of Practice and guidance.*[2]

Emergency procedures relating to lead

346   An accident, incident or emergency, for the purpose of regulation 12, is any event which causes, or threatens to cause, any employee to be exposed to lead or a lead compound on a scale or to an extent well beyond that associated with normal day-to-day activity. For example, any one of the following events may be sufficient to trigger the emergency actions considered by this regulation:

(a)   any serious process fire which could give rise to a serious risk to health;

(b)   any serious spillage of molten lead, uncontrolled release of lead dust or fume, or spillage of lead alkyls liable to make contact with employees' skin;

(c)   any acute process failure that could result in a sudden release of lead fume, vapour or dust; or

(d)   any threatened significant excursion over an OEL for lead, eg where the excursion is clearly the result of an unusual, sudden and serious failure of LEV or other controls other than the reasons for exceeding the limit described in paragraph 115.

347   Accidents and emergencies vary in size and severity, and so whether or not an uncontrolled release or a leak or spillage should be regarded as an incident for the purpose of regulation 12 will depend on the scale of the release and the type and form of the lead concerned. Employers must use their judgement to decide whether the incident can be dealt with under the prevention and control requirements of regulation 6, or whether it is necessary to invoke emergency arrangements drawn up under the Management Regulations supported, where appropriate, by those prepared under regulation 12 of CLAW. The response to an emergency should also be proportionate to the risk, so if employers invoke regulation 12, they should also decide what proportionate action is needed to deal with the situation, eg not all incidents will automatically require the evacuation of the workplace.

348   In the context of these Regulations, 'relevant safety drills' can be any one or more of a number of emergency procedures unique to the circumstances of the particular workplace and incident. It can mean for example: a complete evacuation of the premises; the action taken by certain personnel in the event

of an emergency such as isolating plant or equipment; the steps taken by nominated personnel to help disabled staff leave the building; or a general fire drill. The drills can be practised separately or simultaneously.

349 Employers need not extend the scope of their general emergency procedures drawn up under the Management Regulations if they are satisfied that:

(a) the quantity, type and form of lead at the workplace would create no more than a slight risk because it has a low toxic effect, does not pose a hazard, and exposure to it would not cause any immediate or short-term adverse health effect; and

(b) existing control measures and emergency arrangements are sufficient to contain and control any risk to health the lead might pose during an emergency, and they are capable of restoring the situation to normal as soon as possible.

350 If the conditions described above do not apply, employers must extend their emergency procedures as required by regulation 12 and ensure they are capable of:

(a) mitigating the effects of an incident caused by or directly affecting lead;

(b) restoring the situation to normal as soon as possible; and

(c) limiting the extent of risks to health of employees and, so far as is reasonably practicable, the health of anyone else likely to be affected by the incident, eg the people living in the neighbourhood.

351 Employers may integrate the emergency procedures they draw up under the Management Regulations and CLAW with those required by other legislation which applies to their workplace, eg the Dangerous Substances and Explosive Atmospheres Regulations 2002.

352 To deal with situations which could present significantly greater risks, employers should extend their emergency procedures to include details of the following:

(a) *the identity of the type and form of lead* present at the workplace, where it is stored, used, processed or produced and an estimate of the total amount at the workplace on an average day;

(b) *the foreseeable types of accidents, incidents or emergencies* which might occur involving lead, and the hazards it could present, eg failure of controls, spills, uncontrolled releases of vapours, dusts or fumes into the workplace, accidents with machinery transporting molten lead in the workplace, leaks, or fire. Where such incidents might occur; what effect they might have; the other areas that might be affected by the incident spreading; and any possible repercussions that might be caused;

(c) *the special arrangements* to deal with an emergency situation not covered by the general procedures; the steps to be taken to mitigate the effects;

(d) *the safety equipment and PPE* to be used in the event of an accident, incident or emergency; where it is stored; who is authorised to use it. Judgements about the type of safety equipment, and PPE (including RPE) to be used should be made with regard to the level and type of risk, and a worst-case estimate of the likely concentration of lead in the air at the workplace;

(e) *first-aid facilities* sufficient to deal with an incident until the emergency services arrive; where the facilities are located and stored; the likely effects on the workforce of the accident, incident or emergency, eg burns, scalds, shock, the effects of smoke inhalation etc. Employers should note that they have duties to provide first-aid facilities under the Health and Safety (First Aid) Regulations 1981;

(f) *the role, responsibilities and authority* of the people nominated to manage the accident, incident or emergency and the individuals with specific duties in the event of an incident, eg the people responsible for checking that specific areas have been evacuated, shutting down plant that might otherwise compound the danger, contacting and liaising with the emergency services on their arrival, and making sure that they are aware of the presence of lead and how it might affect the incident;

(g) *procedures for employees* to follow and who should know: how they should respond to an incident and what action they should take; the people who have been assigned specific responsibilities and their roles;

(h) *procedures for clearing up and safely disposing* of any lead damaged or 'contaminated' during the incident;

(i) *regular safety drills;* the frequency of practising emergency procedures will depend on the complexity of the layout of the workplace, the activities carried out, the level of risk, the size of the workforce, the amount of lead involved, the success of each test; and

(j) *the special needs of any disabled employees,* eg assigning other employees to help them leave the workplace in an emergency.

353　The extended procedures should be compiled in consultation with safety representatives, employees and with those people assigned roles and responsibilities during any emergency.

Suitable warning and communication systems

354　Employers must provide suitable communication systems for warning employees who are liable to be affected by an accident, incident or emergency involving lead. The employer may consider it appropriate to provide warning signals for different purposes, ie one type of alarm to warn employees of the need to be prepared to evacuate because an incident is declared, and another signalling the immediate need to evacuate the premises. The communication system the employer provides will be proportionate to the size of the workplace and

91

workforce, the quantity of lead at the workplace, and the level and type of risk it presents. Suitable warning systems might comprise:

(a)   a continuous or intermittent ringing bell, whistle or hooter;

(b)   warning lights;

(c)   intercom, loudspeakers or a public address system.

355   Employers must ensure that all warning systems can be heard/seen in all parts of the premises, and in particular by employees who may work in noisier areas. Employers should also ensure that they take due account of the special needs of disabled employees.

Reviewing the emergency procedures

356   The employer should review, update and replace the emergency procedures in the light of changing circumstances, eg a significant increase in the amount of lead or compound of lead used or processed, changes in the workplace activities involving the use of lead etc.

Making procedures available to the emergency services

357   Employers should ensure that copies of their emergency arrangements and procedures are made available to the relevant internal and external accident and emergency services.

Internal emergency services

358   Internal services include those people assigned specific duties in the event of an accident, incident or emergency, eg people responsible for closing down processes or activities where safe to do so, or liaising with the emergency services on their arrival at the workplace, safety representatives, first aiders etc. Employers should arrange for all the people concerned to be provided with their own copy of the emergency procedures. Copies may be provided on paper or electronically.

359   Copies of the procedures should be circulated and seen by all employees at least once every six months.

External emergency services

360   Employers who extend their emergency procedures to cover situations relating to lead should consider whether it is necessary to make all branches of the emergency service aware of their arrangements. As a minimum requirement, the employer should contact their local fire service and offer to make available a copy of the business's emergency procedures and the collated information on which they are based.

361   The employer's procedures, including details of the amount and type of lead or lead compound present at the workplace, will help the fire service to prepare its own response procedures and precautionary measures in the event of an emergency being declared at the employer's workplace. These measures will ensure that the fire service deals with any declared incident effectively, and especially those likely to occur outside normal working hours, in a way that presents the minimum risk to their own staff.

362 If an incident could have serious repercussions on the environment, the employer should consider offering to make a copy of the business's emergency procedures available to the nearest office of the Environment Agency.

Records

363 A record of the procedures may be kept in writing or recorded by other means, eg electronically. It must be kept readily accessible and retrievable for examination at any reasonable time, eg by safety representative, inspector, etc.

Displaying emergency procedures

364 Where it is practical to do so, employers should consider whether it is appropriate to display the emergency procedures in a prominent position at the workplace for employees to read, eg on employee noticeboard(s). It will be appropriate, for example, where:

(a) the company is fairly small and employees are encouraged to consult their notice board(s) frequently for information about the business and its activities;

(b) the emergency procedures are reasonably short and simple, can be read easily and quickly and can comfortably fit on the notice board.

Employer's action during an emergency

365 The specific actions an employer must take if an accident, incident or emergency occurs are set out in regulation 12(3). Where the incident involves the uncontrolled release of lead dust, fume, vapour or molten lead into the workplace, the employer must exclude all people not concerned with the emergency action from the area of contamination. The employer must ensure that those employees given the task of identifying the source of the release and making repairs, wear appropriate PPE, including where necessary, suitable RPE and protective clothing with which they have been provided until the situation is restored to normal. Depending on the circumstances of the incident, this may mean that employees working in hot contaminated conditions will need frequent rest periods and the opportunity to drink to replace lost fluids.

366 As well as telling employees the cause of the incident and the measures taken or to be taken to resolve it, the employer should also ensure that:

(a) any important lessons learned from it are passed onto the employees and/or their appointed safety representatives; and

(b) ensure that the information is used in any subsequent review of the risk assessment for the process or activity concerned.

367 When an incident is declared, employers also have a duty to tell, and if necessary evacuate, other people who are present in the workplace and who may be affected by it, eg visitors, employees of another employer etc.

Regulation 13

Regulation

13

Exemption certificates

(1)   Subject to paragraph (2), the Executive may, by a certificate in writing, exempt any person or class of persons from all or any of the requirements or prohibitions imposed by Regulations 4, 7, 8, 9(2) and (3) and 10(7) and (11) to (15) of these Regulations and any such exemption may be granted subject to conditions and to a limit of time and may be revoked by a certificate in writing at any time.

(2)   The Executive shall not grant any such exemption unless having regard to the circumstances of the case and, in particular, to -

(a)   the conditions, if any, which it proposes to attach to the exemption; and

(b)   any requirements imposed by or under any enactments which apply to the case,

it is satisfied that the health and safety of persons who are likely to be affected by the exemption will not be prejudiced in consequence of it.

Regulation 14

Regulation

14

Extension outside Great Britain

These Regulations shall apply to and in relation to any activity outside Great Britain to which sections 1 to 59 and 80 to 82 of the Health and Safety at Work etc. Act 1974 apply by virtue of the Health and Safety at Work etc. Act 1974 (Application outside Great Britain) Order 2001[(a)] as those provisions apply within Great Britain.

Regulation 15

Regulation

15

Revocation and savings

(1)   The Control of Lead at Work Regulations 1998[(b)] are revoked.

(2)   Any record required to be kept under the Regulations revoked by paragraph (1) shall, notwithstanding that revocation, be kept in the same manner and for the same period as specified in those Regulations as if these Regulations had not been made, except that the Executive may approve the keeping of records at a place or in a form other than at the place where, or in the form in which, records were required to be kept under the Regulations so revoked.

(a) SI 2001/2127.
(b) SI 1998/543.

Activities in which the employment of young persons and women of reproductive capacity is prohibited

Regulation 4(2)

1.   In lead smelting and refining processes -

(a)   *work involving the handling, treatment, sintering, smelting or refining of ores or materials containing not less than 5 per cent lead; and*

(b)   *the cleaning of any place where any of the above processes are carried out.*

2.   *In lead-acid battery manufacturing processes -*

(a)   *the manipulation of lead oxides;*

(b)   *mixing or pasting in connection with the manufacture or repair of lead-acid batteries;*

(c)   *the melting or casting of lead;*

(d)   *the trimming, abrading or cutting of pasted plates in connection with the manufacture or repair of lead-acid batteries; and*

(e)   *the cleaning of any place where any of the above processes are carried out.*

3.   *In this Schedule, "lead oxides" means powdered lead oxides in the form of lead, lead monoxide, lead dioxide, red lead or any combination of lead used in oxide manufacture or lead-acid battery pasting processes.*

Legislation concerned with the labelling of containers and pipes

Regulation 11(5)

The Chemicals (Hazard Information and Packaging for Supply) Regulations 2002 (CHIP) (SI 2002/1689);

The Health and Safety (Safety Signs and Signals) Regulations 1996 (SI 1996/341);

The Radioactive Material (Road Transport) Regulations 2002 (SI 2002/1093);

The Carriage of Dangerous Goods by Rail Regulations 1996 (SI 1996/2089);

The Packaging, Labelling and Carriage of Radioactive Material by Rail Regulations 2002 (SI 2002/2099);

The Carriage of Dangerous Goods (Classification, Packaging and Labelling) and Use of Transportable Pressure Receptacles Regulations 1996 (SI 1996/2092);

The Carriage of Explosives by Road Regulations 1996 (SI 1996/2093);

The Carriage of Dangerous Goods by Road Regulations 1996 (SI 1996/2095); and

The Good Laboratory Practice Regulations 1999 (SI 1999/3106).

Leadless glaze and the definition of a low-solubility inorganic lead compound

(see paragraph 29 of main text)

Leadless glaze

1   The definition of a 'leadless glaze' was introduced into regulation 3 of the Control of Lead at Work Regulations 1980 (CLAW 1980) by the Potteries etc (Modifications) Regulations 1990 as:

"leadless glaze" means a glaze which does not contain more than one per cent of its dry weight of a lead compound calculated as lead monoxide.

2   However, the subsequently introduced Chemicals (Hazard Information and Packaging for Supply) Regulations 1994 - CHIP 2 (revoked and replaced by CHIP 2002) - required that lead compounds and preparations containing lead compounds, which themselves contain 0.5% or more of lead calculated as the percentage of the element lead in the total weight of the compound or preparation, be classified and labelled as toxic for reproduction. This variance between the two definitions meant that it was possible for a glaze to qualify for definition as leadless under CLAW 1980 but that it could be classified and labelled as toxic for reproduction under CHIP 2. To avoid possible conflict, the provisions of regulation 2 of CLAW 1998 brought the definition used by CLAW of a 'leadless glaze' into line with CHIP 2, and those provisions of CLAW 1998 have been carried over into CLAW 2002. It should be noted, however, that lead frits used in glazes are substances in their own right, and do not come under the generic Approved Supply List entry for 'lead compounds not elsewhere specified' and may therefore be classified differently.

Definition of a low-solubility inorganic lead compound

3   A low-solubility inorganic lead compound is a compound which does not yield to dilute hydrochloric acid more than 5% of its dry weight as soluble lead compound, calculated as lead monoxide, when determined in the manner described in the standard test below. If a lead compound is dispersed in a liquid, eg pottery glaze, then the solid matter should be separated out by a suitable method, eg centrifuging, before applying the standard test and the results should be reported as a percentage of soluble lead, as lead monoxide, in the solid material. The standard test is as follows.

4   A weighed quantity of the material in the form in which it is used or processed which has been dried at 105 ± 2°C thoroughly mixed and completely passed through a sieve of 500 μm aperture size with a minimum of force is to be continuously stirred for 60 ± 1 minutes at 23 ± 2°C with 1000 times its mass of 0.07 M hydrochloric acid. The pH of the hydrochloric acid should be monitored and maintained at its starting value by the addition of (1 + 1) hydrochloric acid.★ The solution should thereafter be allowed to stand for 60 ± 1 minutes at 23 ± 2°C and then filtered before being analysed for lead by means of a suitable analytical technique such as atomic absorption spectrometry. This analysis should be carried out as soon as possible after the preparation of the extract and in any case within 4 hours.

★ Hydrochloric acid (1+1) is prepared by diluting one part by volume of hydrochloric acid, about 36% (m/m), density approximately 1.18 g/ml, with one part by volume of water. In preparing the mixture, the acid should be added to the water (not the water to the acid) to help prevent the mixture producing a violent reaction.

5   In the case of liquid paint analysis it is required to follow the procedure which is described in more detail in BS 3900: Part B3: 1983 *Methods of test for paints. Tests involving chemical examination of liquid paints and dried paint films. Determination of 'soluble' lead in the solid matter in liquid paints: Methods for use in conjunction with the Control of Lead at Work Regulations.*[19] This may also be referred to for general guidance on methodology for the analysis of materials other than liquid paints.

Appendix 2

Technical specifications of the equipment to be used for air monitoring
(see paragraph 219 of main text)

1   The equipment used shall comply with the technical specifications listed below:

(a)   **Sampler characteristics:** use a sampler that is designed to collect the inhalable fraction of airborne particles, as defined in BS EN 481 *Workplace atmospheres. Size fraction definitions for measurement of airborne particles*[13] and that complies with the provisions of BS EN 13205 *Workplace atmospheres. Assessment of performance of instruments for measurement of airborne particle concentrations.*[20]

Note: Existing users may continue to use the single-hole sampler recommended for sampling lead-in-air in previous HSE guidance, since field comparisons have established that it usually gives equivalent results to inhalable samplers for this application.

(b)   **Air-flow rate:** use the samplers at their design flow rate, and in accordance with the instructions provided by the manufacturer, so that they collect the intended fraction of airborne particles.

(c)   **Filter or intake orifice position:** as far as possible, kept parallel to the face of the worker during the whole sampling period.

(d)   **Filter efficiency:** use a filter having a retentivity of not less than 99.5% for particles with a 0.3 µm diffusion diameter.

(e)   **Sampling pumps:** use a sampling pump that is compatible with the samplers used that complies with the provisions of BS EN 1232 *Workplace atmospheres. Pumps for personal sampling of chemical agents. Requirements and test methods.*[21]

2   The lead-in-air sample collected using the equipment specified in paragraph 1(a)-(e) is to be analysed by atomic absorption spectroscopy, inductively coupled plasma atomic emission spectrometry, inductively coupled plasma mass spectrometry or any other technique that gives equivalent results. A suitable procedure is described in MDHS6/3.[14]

Calculation of 8-hour time-weighted averages
(see paragraphs 252-255 of main text)

The 8-hour reference period

1   The term '8-hour reference period' which is used in the definition of 'occupational exposure limit for lead' in regulation 2 of the Control of Lead at Work Regulations 2002 relates to the procedure whereby the occupational exposures in any 24-hour period are treated as equivalent to a single uniform exposure for 8 hours (the 8-hour time-weighted average (TWA) exposure).

2   The 8-hour TWA may be represented mathematically by:

$$\frac{C_1 T_1 + C_2 T_2 + \dots C_n T_n}{8}$$

where C_1 is the occupational exposure and T_1 is the associated exposure time in hours in any 24-hour period.

Averages from full shift samples

3   Results of air monitoring should normally be reported as an 8-hour TWA. If one whole shift sample is taken, the results should be calculated as follows:

(a)   if the working shift is exactly 8 hours, the result of a whole shift sample is the 8-hour TWA;

(b)   if the working shift is less than 8 hours, the TWA can be calculated by assuming a zero exposure during the remaining time.

Example 1

4   An operator works for 7 hours 20 mins on a lead process. The average exposure to lead (other than lead alkyls) during that period is measured as 0.12 mg/m^3.

The 8-hour TWA therefore is

7 hours 20 mins (7.33 h) at 0.12 mg/m^3 and

40 mins (0.67 h) at 0 mg/m^3

That is:

$$\frac{(0.12 \times 7.33) + (0 \times 0.67)}{8} = 0.11 \text{ mg/m}^3$$

This exposure to lead is significant because it exceeds half the occupational exposure limit for lead but is below the limit itself.

5   If the working shift is more than 8 hours, a TWA should be calculated as a representative 8-hour period of the working day. If the exposure to lead over the working week is longer than 40 hours, this should be taken into account when interpreting the results, preferably by an experienced occupational hygienist.

6   The following example is a guideline for the interpretation of results when an employee has been exposed to lead for a 50-hour working week. It

98

shows the 8-hour TWA being increased on a *pro rata* basis for comparison with the occupational exposure limit for lead.

Example 2

7   An employer works a 10-hour shift for 5 days. An analysis of the 8-hour shift sampling data shows a lead-in-air exposure of 0.07 mg/m^3. As a guideline for estimating the exposure of the employee working 50 hours each week, the 8-hour TWA is increased *pro rata* as follows:

$$\frac{0.07 \times 50}{40} = 0.088 \text{ mg/m}^3$$

The employee's exposure is significant after adjustment for the longer working week.

Averages from split samples

Example 3

8   Working periods may be split into several sessions for the purpose of sampling to take account of rest and meal breaks etc. This is illustrated by the following work pattern:

Working period	Sampling result	Duration of sampling(h)
08.00-10.30	0.12	2.5
10.45-12.45	0.07	2.0
13.30-15.30	0.20	2.0
15.45-17.15	0.10	1.5

$$\text{8-hour TWA} = \frac{2.5(0.12) + 2(0.07) + 2(0.20) + 1.5(0.10)}{8}$$

$$= \frac{0.30 + 0.14 + 0.15}{8} = 0.12 \text{ mg/m}^3$$

The exposure is significant but below the occupational exposure limit for lead.

The results for individual periods show high exposure during one working period (13.30-15.30) and this should be investigated for possible changes in working patterns or processes with corrective action taken in line with regulation 6 of CLAW. This valuable information would not have been apparent from full shift sampling.

Appendix 4   Methods of measuring blood-lead and urinary lead concentrations and other biological indicators
(see paragraphs 262, 274 and 276 and categories B and C of Table 3 in main text)

Blood lead: atomic absorption spectroscopy or equivalent method.

ALAU: Davis[22] or equivalent method.

ZPP: Haematofluorimetry[23] or equivalent method.

ALAD: European standardised method[24, 25] or equivalent method.

99

Guidance notes for appointed doctors on the Control of Lead at Work Regulations 2002
(see paragraph 269 of main text)

Introduction

1   The Regulations aim to protect the health of people at work by preventing or, where this is not reasonably practicable, adequately controlling their exposure to lead. If exposure to lead is significant (as defined by the Regulations) the amount of lead employees absorb needs to be monitored. Individuals can then be removed from further work with lead before their health is affected.

2   Comprehensive guidance on the Regulations is given in the main text. It is of paramount importance that this is read and adhered to by appointed doctors. The guidance notes produced here provide information on lead exposure, toxicity and clinical effects of lead poisoning. They give guidance to appointed doctors on biological monitoring and medical surveillance under the Control of Lead at Work Regulations 2002 (regulation 10).

Exposure to lead

3   Lead and its compounds are used or encountered in a wide variety of industries. These include: mining, smelting, refining, alloying and casting; lead-acid battery manufacture; jewellery manufacture; leaded-glass manufacture; manufacture of pigments and colours; glazes and transfers in the pottery industry; manufacture of inorganic and organic lead compounds; electronics industry; antique restoration (stripped pine); shipbuilding, repairing and breaking; demolition industry; some painting of buildings; some spray painting of vehicles; scrap industry; vehicle radiator repair; stabilising some plastics in the manufacture of pipes; use of lead alloys, eg manufacture of solder; and on firing ranges. Two lead alkyls (tetraethyl and tetramethyl lead) are manufactured in Britain and are used as antiknock agents in petrol. However, since the introduction of unleaded petrol the use of antiknock compounds based on lead alkyls has fallen significantly. However, there continues to be the potential for exposure when tanks, which have contained such compounds, are cleaned.

Lead and lead compounds, except lead alkyls

4   Exposure to lead and its compounds, except lead alkyls, occurs by two routes: inhalation and ingestion. In different industries and also in different processes in the same industry, either of these can be the major route of entry into the body. Lead can be present in two forms: fume, generated at temperatures greater than 500°C, and dust. Occupational exposure to lead is dependent not only on the concentrations of lead in workplace air but also on the personal hygiene and personal habits of the worker.

5   Any process involving lead which leads to the production of dust or fume can result in inhalation of lead, eg cutting, grinding and burning. However, every process/job involving lead (and its compounds) in any way can lead to exposure by ingestion.

Lead alkyls

6   Exposure to lead alkyls can occur by three routes: inhalation, ingestion and absorption through the skin. Exposures can occur during manufacture of the compounds as well as during maintenance and cleaning of petrol storage tanks.

Control of exposure

7   For many processes no single method of control will be sufficient to adequately control exposure and a combination of measures may be necessary. Exposure to lead should be reduced where possible by use of appropriate engineering controls, personal protective equipment, good housekeeping, adequate washing facilities, safe systems of work and provision and maintenance of respiratory protective equipment. Even if all this has been provided, the attitude of an individual to the risks involved can significantly affect exposure.

Lead toxicity and clinical effects of lead poisoning

Lead toxicity

Lead and lead compounds, except lead alkyls

8   Lead is known to produce a continuum of diverse biological effects in humans, depending on the dose, which are usually associated with long-term exposure. These range from minor biochemical changes, which are unlikely to have adverse health consequences, to severe irreversible or life-threatening disruption of a number of organ systems, in particular the haemopoietic, nervous and renal systems. The relationship between blood-lead concentration and hypertension remains contentious, the balance of argument suggesting little effect of lead on blood pressure. There are also concerns about the possible effects of lead on the reproductive system and developing foetus, although the evidence is less clear.

9   Lead has been shown to inhibit enzymes involved in haem synthesis, resulting in a decrease in its production and the accumulation of δ-aminolaevulinic acid, zinc protoporphyrin (ZPP) and coproporphyrin. Changes in related biochemical parameters can be detected at blood-lead concentrations below 40 μg/dl and there appears to be an increased risk of anaemia at concentrations above 50 μg/dl. The blood film in lead-induced anaemia shows punctate basophilia. High levels of exposure to lead have historically been associated with both central and peripheral nervous system toxicity and kidney damage, often accompanied by abdominal colic. However, serious manifestations of neuro- and renal toxicity are unlikely to occur at blood-lead concentrations below 100 μg/dl.

10   Concern for possible effects on male fertility stems from a number of studies which looked for effects on spermatogenesis or endocrine disturbances in lead workers. Firm conclusions cannot be drawn from these studies but there is limited evidence of an association between reduced semen quality and blood-lead concentrations in excess of 40 μg/dl.

11   There are reports from the early 1900s of an association between occupational lead exposure and female infertility, miscarriage and stillbirth. At this time lead exposure was poorly controlled and exposure levels were considerably higher than encountered in modern industry. The influence of pre- and postnatal low-level environmental lead exposure on *in utero* and childhood development has been the subject of intense research activity in recent years and is a controversial area. These studies have shown no evidence of an association between lead exposure and spontaneous abortions or birth defects. In several studies a correlation between lead exposure and reduced length of gestation and with reduced birth weight has been reported in association with maternal blood-lead concentrations as low as 20 μg/dl. Some studies have also suggested that there may be an association between prenatal or perinatal lead exposure and performance deficits in tests of postnatal

neurological and psychomotor development. However, from the data available on these associations, it cannot be concluded with certainty, at the present time, that there is a causal relationship with lead exposure.

Lead alkyls

12 The available human data are limited. There are a number of human cases with neurological and psychiatric symptoms, which can progress to coma and death, but little is known about the dose-response relationship. These have resulted from acute or short-term repeated overexposure, usually by the inhalation route. In animal experiments the main effect of repeated exposure is also neurotoxicity, including irritability, tremors, ataxia or coma, and pathological changes such as: oedema and degenerative lesions in the brain and spinal cord; peripheral nerve damage.

Clinical effects of lead poisoning

Diagnosis

13 Biological monitoring of lead workers can only indicate those who are at risk of lead poisoning. Risk of poisoning increases with absorption of lead and lead compounds. However, there may be marked individual variation. Diagnosis rests with clinical judgement and the term 'lead poisoning' should only be applied where relevant symptoms and clinical signs have been identified. The diagnosis should be supported by biological monitoring data or an adequate history of lead exposure.

Lead and lead compounds, except lead alkyls

14 Lead poisoning may develop rapidly but is more often insidious in onset. Clinical illnesses include anaemia, acute abdominal colic, acute and chronic encephalopathy, peripheral neuropathy and chronic lead nephropathy. However, initial symptoms may be non-specific, the most important being fatigue and lassitude which may destroy the social life of the affected individual. Anorexia, headache, arthralgia, facial pallor (which may be unrelated to anaemia), dyspepsia, constipation or intermittent diarrhoea and a metallic taste in the mouth, may also occur. During the later stages of lead poisoning there may be abdominal discomfort, colic, vomiting and weakness of those muscle groups most often used. A 'Burtonian' or blue line may appear at the gum margin. This is associated with poor oral hygiene and indicates lead absorption but is not diagnostic of lead poisoning.

15 An individual with poisoning should be removed from further exposure to lead. Chelation therapy, carried out under hospital in-patient supervision, may be warranted.

Lead alkyls

16 The principal feature of poisoning by lead alkyls is a rapidly developing and life-threatening encephalopathy. The individual might be irritable, restless and confused, perhaps intermittently at first. Ataxia and tremor may occur. As the condition progresses, the individual becomes increasingly disorientated and unco-operative. There may also be nausea, abdominal pain and vomiting. Extreme cases can become overtly psychotic and self-destructive or may lapse into a coma with subsequent convulsions or cardiorespiratory failure.

17 Lead alkyl poisoning is a medical emergency requiring immediate admission to hospital. Receiving doctors should be aware of the diagnosis. Use

of chelating agents appears to be of little benefit. Supportive therapy, including sedation and life support, is the treatment of choice.

Notification of lead poisoning

18   Lead poisoning is reportable under the Reporting of Injuries, Diseases and Dangerous Occurrences Regulations (RIDDOR).[26] The employer has a duty to report cases to the relevant authority. It is also a prescribed disease. The affected person has a right to claim benefit from the Department for Work and Pensions. Claim forms are available from local offices. Completed forms should be supported by a letter from the appointed doctor.

Biological monitoring

19   Biological monitoring is a tool for assessing the uptake of lead. It may also help in evaluating health risks associated with exposure.

20   The Control of Lead at Work Regulations 2002 require that medical surveillance, including biological monitoring, be carried out on employees where appropriate (see paragraphs 262-321 of the main text and paragraphs 27-54 of this appendix).

21   In the workplace there is evidence of a relationship between blood-lead levels and air levels. Blood-lead levels show a general increase with increasing air levels. However, this relationship is so weak that, for an individual worker, measurements of personal exposure are a poor predictor of their blood-lead level. Because this relationship is so weak, and substantial differences in the relationship have been observed between different workplaces, air levels are unlikely to be the only determinant of blood-lead. Other factors likely to influence blood-lead include physical characteristics of the form of lead, namely particle size and solubility, oral ingestion related to poor personal hygiene or housekeeping practices and lead exposures outside work.

Lead and lead compounds, except lead alkyls

22   Once absorbed, the toxicokinetic properties of lead and its compounds, except lead alkyls, are assumed to be similar. Absorbed lead is transported by the blood, mainly in the erythrocytes. Distribution around the body is widespread although not homogeneous, with approximately 94% of the total body burden of adults being located in bone. Normal bone turnover provides a continuing input of this accumulated lead to the blood. The level of input depends on the extent of lead stored in the bone. In conditions of increased bone turnover, for example in pregnancy and during breastfeeding, increased mobilisation of lead from bone is of concern. No information is available on the metabolism of lead. In humans, elimination half-lives for lead in blood and soft tissues have been estimated to be about 30-40 days. For bone, the half-life is likely to be in excess of 20 years. Lead can cross the placenta and can be transferred to breast milk from the maternal circulation. Absorbed lead is eliminated primarily in urine, although excretion via bile is also an important route. Other more minor excretory routes include sweat, hair and nails.

23   For biological monitoring of occupational exposure to lead and its compounds, except lead alkyls, blood-lead concentration is the most commonly used indicator. Interpretation of blood-lead concentration should take into account that it mainly represents the most recent level of exposure. However, in active lead workers with low current exposure, blood-lead levels may be influenced by normal mobilisation of lead pools from the skeleton. The level of skeleton lead burden or body burden is determined by cumulative exposure over years.

24   At the discretion of the appointed doctor, other relevant biological tests may be carried out as part of the periodic medical assessment (see paragraph 274 of the main text). Measurement of haemoglobin concentration is a non-specific test as there are many causes of anaemia. Therefore results should be evaluated carefully. Measurement of ZPP levels is considered to be the most specific and sensitive indicator of lead-induced impairment of haem synthesis. The only other condition associated with a raised ZPP is iron deficiency anaemia. Levels in unexposed workers are usually less than 2 µg/g haemoglobin. On average, ZPP levels begin to rise when blood-lead levels reach 30 µg/dl.

Lead alkyls

25   Lead alkyls are readily absorbed from the respiratory and gastrointestinal tracts and through the skin. Once absorbed, lead alkyls and their metabolites are transported in the blood and become widely distributed throughout the body. Highest concentrations are found in the liver, followed by the kidney, brain and muscle. It has been shown that lead alkyls can cross the placenta and it is anticipated that transfer to breast milk will occur. The primary routes of excretion are by exhalation of volatile organo-lead species and via urine and faeces.

26   The most common method of biological monitoring for occupational exposure to lead alkyls involves measurement of total lead in urine. Total lead in urine reflects exposure to both inorganic lead and lead alkyls.

Medical surveillance

27   The general requirements for medical surveillance are covered in paragraphs 262-321 of the main text. Employees should be under medical surveillance as required by regulation 10(1) where exposure to lead is significant as defined in regulation 2, or in accordance with the guidance in paragraphs 263 and 266 of the main text. Suitable medical surveillance procedures consist of initial and periodic clinical assessments and biological monitoring.

28   The objectives of medical surveillance are to:

(a)   make an initial assessment of the suitability of an employee to work with lead;

(b)   evaluate the effect of lead absorbed by an employee and advise on state of health;

(c)   monitor exposure of female employees of reproductive capacity and advise on the special need to protect any developing foetus;

(d)   assess the suitability of an employee to continue in work where there is continuing exposure to lead;

(e)   detect early health effects of excessive lead absorption and remove employees from exposure to prevent development of further health effects; and

(f)   assist employers in their duty to control exposure of employees to lead.

29   A 'woman of reproductive capacity' may be considered as any woman who is physiologically and anatomically capable of conceiving. Capacity to conceive is the only test that should be applied. Matters such as marital status,

sexual orientation, cohabitation with a sterile man and use of contraception, are all irrelevant. A 'young person' is defined as an individual who has not attained the age of 18 years and is not a woman of reproductive capacity.

Lead and lead compounds, except lead alkyls

Initial medical assessment

30   The initial medical assessment should be carried out before starting work with lead, and certainly within 14 working days of that date. At this stage it is necessary to know the type(s) of lead to which the employee will be exposed. If the individual has been exposed to lead at work and under medical surveillance in the previous three months, the former employer may have provided the individual with a copy of relevant medical surveillance records. The appointed doctor should then ask to see the employee's copy and request permission to duplicate any appropriate information. If the new employee has no copy of previous medical records, the appointed doctor should ask the individual's permission to approach the former employer to request a copy of these records. Where a new employee has been exposed to lead at work more than three months previously, the appointed doctor may use discretion in attempting to obtain medical records.

31   Consideration should be given to occupational record, medical history, personal hygiene and intellectual capacity to work with hazardous substances. Habits such as cigarette smoking and nail biting may be particularly relevant. Workers with poor personal hygiene are more likely to absorb excessive amounts of lead. A clinical examination must be conducted, noting problems such as extensive dermatitis. For individuals suffering from anaemia, especially due to iron deficiency, consideration should be given to exclusion from work with lead until the condition is diagnosed and treated.

32   Samples should be taken for measurement of baseline blood-lead and haemoglobin levels. Further biological tests may be used to verify results of baseline measurements. A list of such tests is given in Appendix 4 of this publication.

33   At initial examination, form FOD MS98 (Annex 1) should be completed for each worker and written consent obtained, if possible, for the disclosure of biological test results to the employer. Each worker should be given a copy of HSE's free leaflet *Lead and you.*[17]

34   If the clinical examination and test results are satisfactory, written confirmation of the assessment that the employee is fit to work with lead must be entered in the appropriate health record. Arrangements should be made for the next examination and, where appropriate, for measuring the employee's blood-lead, ie within three months or earlier of the initial measurement (see paragraph 277 of the main text) and the employer notified. It is then the duty of the employer to ensure workers present themselves at the specified time. Arrangements should be made to see absentees on a convenient day after their return to work. Unresolved difficulties experienced with employees failing to attend for repeat examinations should be referred to the senior medical inspector (formerly known as a senior employment medical adviser).

Periodic medical assessments

35   Periodic medical assessments should consist of measurement of blood-lead concentrations and, at the discretion of the appointed doctor, other relevant biological tests. These may include at least yearly measurement of

105

haemoglobin and ZPP levels. A clinical assessment should be conducted at least once a year. This should incorporate a review of occupational history, medical records and physical examination. The information should be entered onto form FOD MS99 (Annex 2). Observed patterns of blood-lead levels and their implications for work practices, personal hygiene, change in exposure and current ill health, require consideration.

36   Intervals between periodic medical assessments should not exceed 12 months. The appointed doctor may decide the frequency of periodic medical assessments where blood-lead concentration is below the appropriate suspension level.

37   Blood-lead should be measured every three months if employees are significantly exposed to metallic lead and its compounds. If exposure is uniform and a consistent blood-lead pattern is established, frequency of measurements can be reduced in accordance with paragraph 278 of the main text. Some exposures, such as burning lead paint, may be so variable that a clear pattern of lead absorption cannot be established. Under such circumstances, it may be necessary to carry out measurements even more frequently, eg every two weeks, depending on the results obtained. Women of reproductive capacity and young persons, with significant exposure to lead, should have their blood-lead levels assessed at least every three months.

Action levels

38   All employees with significant exposure to lead should be subject to action levels. These are based on the blood-lead concentrations given for different groups of employees in paragraphs 279 and 280 of the main text. The action level warns the employer that the blood-lead concentration of an employee is approaching suspension level. To reduce blood-lead concentration below the action level, the employer will need to investigate the effectiveness of control measures.

Suspension levels

39   Blood-lead concentrations at which workers are considered for suspension from further exposure to lead are given in Table 5 in the main text. Where blood-lead concentration reaches the suspension level, an *urgent* confirmatory test should be requested. Every effort should be made to obtain the result of the repeat test *within ten working days* of the initial result becoming available. Where the initial result relates to a young person or woman of reproductive capacity, or it reaches or exceeds 70 µg/dl for all other employees, the doctor should consider whether to remove the employee concerned from work involving exposure to lead until the result of the repeat test becomes available. If the result from the repeat test is equal to or greater than the suspension level, the appointed doctor should suspend the employee from further work involving exposure to lead.

40   Employees, excluding women of reproductive capacity, who have worked with lead for a prolonged period may have a high body burden of lead which takes a long time to fall below the suspension limit. These employees:

(a)   have been exposed to lead at work for at least 20 years; or

(b)   are aged 40 years or more and have been exposed to lead at work for at least 10 years.

41   Such individuals may be further exposed to lead, providing that the blood-lead concentration and results of other biological tests are compatible

with the values specified in paragraph 287 of the main text. The employer should make every effort to reduce the blood-lead level of the employee. The appointed doctor should also consider increasing the frequency of testing for blood-lead and haemoglobin levels.

42   The appointed doctor may use discretion to certify whether an employee is fit for work involving further lead exposure even if the blood-lead concentration is below the suspension level.

Lead alkyls

Medical assessments

43   Biological monitoring for absorption of lead alkyls should be conducted by measuring total urinary lead concentration. The latter is corrected for urinary creatinine concentration to account for differences in the volumes of urine produced over unit of time by the kidney. This measurement should be used for carrying out initial and periodic medical assessments of employees exposed to lead alkyls.

44   The general arrangements for medical surveillance of workers exposed to lead and lead compounds (except lead alkyls) also apply to employees exposed to lead alkyls. Those with a recent history of psychiatric illness should not work with concentrated organic lead compounds, as relapses may be confused with lead poisoning. At least once a year the medical assessment should include measurement of blood-lead as well as urinary lead. A clinical assessment should be conducted where indicated by the results of these tests. The intervals between periodic evaluation of urinary lead are shown for different groups of employees in Table 6 in the main text.

Action levels

45   No action levels have been set for employees exposed to lead alkyls.

Suspension levels

46   Urinary lead concentrations at which employees reach suspension level are given in Table 7 in the main text. If the suspension level is triggered, an *urgent* repeat test should be requested. Every effort should be made to obtain the result of the repeat test *within ten working days* of the initial result becoming available. Where the initial result relates to a woman of reproductive capacity, the doctor should consider whether to remove the employee concerned from work involving exposure to lead alkyls until the result of the repeat test becomes available. If the result from the repeat test is equal to or greater than the suspension level, the appointed doctor should suspend the employee from further work involving exposure to lead alkyls.

47   The appointed doctor may use discretion to certify whether an employee should be removed from work with lead alkyls even if the suspension level has not been reached.

Unsuitability for work with lead

General action

48   When the suspension level for any employee is reached, the appointed doctor should make an entry in the appropriate health record. This states whether an employee should be removed from further exposure to lead or any conditions under which further exposure may continue. The appointed doctor

should then complete and forward a copy of the Certification of fitness/unfitness form (ie form FOD MS103 - see Annex 4) to an appropriate medical inspector (employment medical adviser) within seven days of certifying any employee as unfit. In addition, the appointed doctor should advise the medical inspector of any unsatisfactory trends or matters of concern. If an employee is temporarily removed from work involving exposure to lead, the relevant doctor should seek the employee's consent and write to tell the employee's own doctor (general practitioner) so that the GP can make a note in the employee's medical records. Following suspension, a decision on when an employee can return to work involving exposure to lead should be authorised by the appointed doctor.

Women of reproductive capacity

49   The action and suspension levels for women of reproductive capacity are lower than those for other employees. A woman of reproductive capacity who becomes pregnant should inform the employer at the earliest opportunity. The employer should then notify the appointed doctor. A pregnant employee should be removed from any work where exposure to lead is liable to be significant. These measures are designed to protect the developing foetus. There is no direct evidence to demonstrate that certain blood-lead levels adversely affect the health of a foetus. However, it is known that lead crosses the placenta and there is some limited evidence from human studies and animal experiments that suggests the foetus may be at risk.

Records

50   The employer should ensure that basic details of all those who need to be under medical surveillance are recorded. This information should contain the items summarised in Appendix 6. The initial medical assessment should be recorded on Form FOD MS98 (Annex 1). It includes information on occupational and medical history, clinical examination, biological tests and evaluation of fitness for work with lead. Subsequent periodic medical assessments should be recorded on form FOD MS99 (Annex 2).

51   The appointed doctor should make available to the employer, results of biological tests where employees have given their consent. The employer should always be given sufficient details of results to permit any adverse trends to be investigated and reversed. Form FOD MS102 (Annex 3) can be used for notifying employers of biological test results and to provide a record of medical surveillance. The appointed doctor should immediately inform the employer of any employees whose blood-lead concentrations have reached or exceeded the appropriate action level.

52   The employer should be given a record of all employees who are certified by the appointed doctor for suspension from further work with lead or who may continue working with lead only under specified conditions. Form FOD MS103 (Annex 4) is appropriate for this purpose. It also allows information to be recorded on return to work of previously suspended employees and any conditions that are imposed upon them when they resume work with lead.

53   Medical surveillance records should be retained for a period of at least 40 years.

Statistical returns

54   Once a year the appointed doctor should promptly submit to HSE a statistical return of workers under medical surveillance using the form 'EMSU BLOODLEAD 1' (Annex 5). Guidance on completion of EMSU BLOODLEAD 1 is given with the form.

Form FOD MS98: Initial medical assessment

HSE
Health & Safety
Executive

Employment Medical Advisory Service
Control of Lead at Work Regulations 2002

Initial medical assessment

Name

Permanent address

N.I. number   Date of birth   Sex

Employer's name

Employer's address

Years exposed to lead before starting in current employment	Date of first exposure to lead in current employment	Date of end of exposure to lead in current employment

GP name   GP tel no

GP address

HISTORY

Occupational (including specifications of previous employment involving lead exposure)

Medical (including smoking history)

Clinical examination (including personal hygiene, nail biting etc)

Consent given to disclosure of biological results to employer   Yes   No

Laboratory test results

Name of laboratory [_____]   Included in HSE list   Yes [ ]   No [ ]

Test	Result	Units
Blood Lead		
Haemoglobin		
Urinary Lead		
Other		

Blood Lead range code (A, B, C, D, E - see footnote below)   [ ]

Assessment of fitness   Fit [ ]   Unfit [ ]

Restrictions (if any)

[   ]

Employer informed of result   Yes [ ]   No [ ]

Employee informed of result   Yes [ ]   No [ ]

Date of review   [_____]

Name EMA/AD   [_____]

Signed   [_____]   Date   [_____]

Footnote: Blood-lead (µg/dl) range codes:

A   under 30
B   ≥30 <40
C   ≥40 <50
D   ≥50 <60
E   60 and over

Form FOD MS99: Surveillance record for person exposed to lead

HSE Health & Safety Executive

Employment Medical Advisory Service
Control of Lead at Work Regulations 2002

Surveillance record for person exposed to lead

Medical in Confidence

Name and address of employer

Name

D.O.B.

Consent given to disclosure of biological test results to employer   Yes [ ]   No [ ]
(delete whichever inapplicable)

Included in HSE list   **Yes** [ ]   **No** [ ]

Name of Laboratory [ ]

Assess- ment date	Work activity and reason for surveillance	Sample taken	Results of laboratory analyses					Clinical notes and assessment, including review of Medical History, details of certification of unfitness/ fitness for work involving exposure to lead or other action	Other action	Date for next review	Blood-lead range code- see footnote overleaf
			Blood lead (µg/dl)	Haemo- globin	Urinary lead (units to be specified)	ZPP	Other analyses (units and type to be specified)				
		Blood									
		Urine									
		Blood									
		Urine									
		Blood									
		Urine									
		Blood									
		Urine									

FOD MS99 (10.02)

continued overleaf

111

Assessment date	Work activity and reason for surveillance	Sample taken	Results of laboratory analyses					Clinical notes and assessment, including review of Medical History, details of certification of unfitness/fitness for work involving exposure to lead or other action	Other action	Date for next review	Blood-lead range code - see footnote below
			Blood lead (µg/dl)	Haemo-globin	Urinary lead (units to be specified)	ZPP	Other analyses (units and type to be specified)				
		Blood									
		Urine									
		Blood									
		Urine									
		Blood									
		Urine									
		Blood									
		Urine									
		Blood									
		Urine									

Footnote:

Blood-lead (µg/dl) range codes:

A under 30
B ≥30 <40
C ≥40 <50
D ≥50 <60
E 60 and over

Form FOD MS102: Notification to employer of biological test results and record of medical surveillance

HSE
Health & Safety
Executive

Employment Medical Advisory Service
Control of Lead at Work Regulations 2002

Notification to employer of biological test results and record of medical surveillance

To:

From: (Name and address of EMA/AD)

Date of assessment test	Name	D.O.B.	Work activity	Clinical assessment (Yes or No)	Consent given to disclosure of results to employer (Yes or No)	Biological tests		Date of next assessment
						Type	Result (actual where employee consents; otherwise blood-lead range - see footnote overleaf)	

FOD MS102 (10.02)

| Date of assessment test | Name | D.O.B. | Work activity | Clinical assessment (Yes or No) | Consent given to disclosure of results to employer (Yes or No) | Biological tests | | Date of next assessment |
						Type	Result (actual where employee consents; otherwise blood-lead range - see footnote below)	

Employment Medical Adviser's/Appointed Doctor's signature: Date:

Footnote: Blood-lead (µg/dl) range codes:
A   under 30
B $\geq$30 <40
C $\geq$40 <50
D $\geq$50 <60
E   60 and over

Form FOD MS103: Certification of unfitness/fitness

Employment Medical Advisory Service
The Control of Lead at Work Regulations 2002

HSE Health & Safety Executive

Certificate of unfitness/fitness

Employer's name and address

From (name and address of EMA/AD)

Name of worker	Work activity	Date of medical assessment	Unfit for exposure to lead or unfit for exposure to lead under stated conditions	EMA's/AD's signature and date	Fit to return to work with lead (if subject to conditions please specify)	EMA's/AD's signature and date

Certifications by EMA

FOD MS103 (10.02)

Name of worker	Work activity	Date of medical assessment	Unfit for exposure to lead or unfit for exposure to lead under stated conditions	EMAs/AD's signature and date	Certifications by EMA Fit to return to work with lead (if subject to conditions please specify)	EMA's/AD's signature and date

Form EMSU BLOODLEAD 1 - Control of Lead at Work Regulations 2002: Annual return of persons under medical surveillance

Employment Medical Advisory Service

HSE
Health & Safety Executive

Control of Lead at Work Regulations
Annual return of persons under medical surveillance

PIN No. ☐☐☐☐☐☐

Medical Inspector - Insert 1
Appointed Doctor - Insert 2 ☐

Employer's name   Doctor's name

Address   Address

Post code

Telephone number   Telephone number

Type of business

FOR GUIDANCE ON COMPLETION PLEASE READ NOTES ON PAGE 2

Year ended 31 March   20 ☐☐

1. Lead industry sector code *(see note 1 page 2)*   e.g. 02 ☐☐

2. Highest blood level measured during year
   (see notes 2&3 page 2)

μg / 100ml	Total number of males	Of which under 18yrs (see note 3 p2)	Total number of females	Of which under 18yrs (see note 3 p2)
Under 10				
10 - 19				
20 - 24				
25 - 29				
30 - 34				
35 - 39				
40 - 49				
50 - 59				
60 - 69				
70 - 79				
80 and over				
Total under medical surveillance during year				

3. Total number of persons suspended due to excess blood lead *(see note 4 page 2)*

Total number of males	Of which under 18yrs	Total number of females	Of which under 18yrs

4. Suspensions from lead work *(see note 5 page 2)*

	Total number of males	Of which under 18yrs	Total number of females	Of which under 18yrs
a. Suspensions due to excess blood lead				
b. Suspensions due to other medical reasons				
Total (a + b) suspensions during the year				

5. Total number of blood leads measured during year *(see note 6 page 2)*

Total number of males	Of which under 18yrs	Total number of females	Of which under 18yrs

For office use only

Input by   Identity number ☐☐☐☐☐☐☐☐

Date of input

EMSU BLOODLEAD 1 (rev 10.2002)   1   *delete as appropriate

117

NOTES FOR GUIDANCE ON COMPLETION

This form is to be completed by the doctor responsible for medical surveillance (as required by the Regulations) at 31 March each year taking into account the records left by any other doctor who may have been responsible for surveillance earlier in the year. Any difficulties should be discussed with the local Employment Medical Adviser.

1. Lead industry sector code

01 Smelting, refining, alloying, casting
02 Lead battery industry
03 Badge and jewellery enamelling and other vitreous enamelling
04 Glass making
05 Manufacture of pigments and colours
06 Potteries, glazers and transfers
07 Manufacture of inorganic or organic lead compounds (including lead salts, fatty acids)
08 Shipbuilding, repairing and breaking
09 Demolition industry
10 Painting buildings and vehicles
11 Work with metallic lead and lead containing alloys
12 Other processes
13 Scrap industry

These should already be shown on medical surveillance record sheets. Where more than one industry code is applicable within the same premises, a separate form should be completed for each industry.

2. Highest blood lead measured during year

This table should show for each of the blood lead ranges the total numbers of males and females whose highest blood lead level measurement for the year lie within those ranges. **Each person should be counted once and once only in the category corresponding to their highest measured level.**

The total under surveillance in the year is the total of the eleven ranges and represents the total numbers of males and females whose blood lead levels have been measured.

3. Young people

For the purpose of this form under 18 year olds are defined as those who are aged under 18 years for the *majority* of the operating year i.e. whose birthday comes on or after 1 October. e.g. Year ended 31 March 2003: under 18 if date of birth 01.10.84 **or after.**

4. Individuals suspended

Enter the total numbers of persons certified as unfit for work due to excess blood lead. If a person was certified unfit on more than one occasion, count them once only in this table.

Certificates of unfitness continuing beyond 31 March should for the purposes of items 3 and 4, be counted in the year of issue and not repeated in the following year.

5. Suspensions from lead work

Enter the number of certificates of unfitness issued as distinct from the previous number of persons so certified, separate totals being required for certificates issued because of excess blood lead and those issued for other medical reasons. **If a worker has been certified as unfit on two occasions, each certificate should be shown in its appropriate category with both counting towards the total.**

Certificates of unfitness continuing beyond 31 March should, for the purposes of items 3 and 4, be counted in the year of issue and not repeated in the following year.

6. Blood leads measured

Enter the total number of blood lead measurements made during the year for males and females. **If a worker's blood lead was measured more than once, count each measurement.**

2

Recorded details of employee under medical surveillance because of exposure to lead
(see paragraph 309 of main text)

Employee's details

Surname
Forenames
Maiden name (if applicable)
Permanent address
Place of birth (town/city)(county)
Date of birth
Sex
NI number
NHS number (if available)

Doctor's (GP) details

Name
Address
Telephone number
Fax number (if any)

Employer's details

Name
Address
Telephone number
Fax number (if any)

Employment details

Years exposed to lead before starting current employment
Date of first exposure to lead in current employment (day)(month)(year)
Date of end of exposure to lead in current employment (day)(month)(year)

Additional information

(a) the reason for medical surveillance;
(b) the dates of initial and periodic medical surveillance;
(c) the results of clinical assessments;
(d) the results of measuring blood-lead concentrations and of any other biological tests in enough detail to allow adverse trends to be identified; and
(e) action taken, including periods moved to work not involving exposure to lead, and periods of suspension.

References

1   *Workplace health, safety and welfare. The Workplace (Health, Safety and Welfare) Regulations 1992. Approved Code of Practice and guidance* L24 HSE Books 1992 ISBN 0 7176 0413 6

2   *Management of health and safety at work. The Management of Health and Safety at Work Regulations 1999. Approved Code of Practice and guidance* L21 HSE Books 1992 ISBN 0 7176 2488 9

3   *Safety representatives and safety committees* L87 (Third edition) HSE Books 1996 ISBN 0 7176 1220 1

4   *A guide to the Health and Safety (Consultation with Employees) Regulations 1996. Guidance on Regulations* L95 HSE Books 1996 ISBN 0 7176 1234 1

5   *Consulting employees on health and safety: A guide to the law* Leaflet INDG232 HSE Books 1996 (single copy free or priced packs of 15 ISBN 0 7176 1615 0)

6   *Control of substances hazardous to health. The Control of Substances Hazardous to Health Regulations 2002. Approved Code of Practice and guidance* L5 (Fourth edition) HSE Books 2002 ISBN 0 7176 2534 6

7   *The selection, use and maintenance of respiratory protective equipment: A practical guide* HSG53 (Second edition) HSE Books 1998 ISBN 0 7176 1537 5

8   *Fit testing of respiratory protective equipment facepieces* Information Document HSE 282/28 (rev) 2002 Available as pdf file on HSE's website: www.hse.gov.uk

9   *Accreditation for the inspection of local exhaust ventilating (LEV) plant* RG4 The United Kingdom Accreditation Service (UKAS) 2000 (Available from UKAS, 21-47 High St, Feltham, Middlesex TW13 4UN Tel: 020 8917 8421 (9 am-1 pm) Fax: 020 8917 8500 Website: www.ukas.com)

10   *Maintenance, examination and testing of local exhaust ventilation* HSG54 (Second edition) HSE Books 1998 ISBN 0 7176 1485 9

11   BS EN 4275: 1997 *Guide to implementing an effective respiratory protective device programme* British Standards Institution

12   *Occupational exposure limits: Containing the list of maximum exposure limits and occupational exposure standards for use with the Control of Substances Hazardous to Health Regulations 1999* Environmental Hygiene Guidance Note EH40 (revised annually) HSE Books 2002 ISBN 0 7176 2083 2

13   BS EN 481: 1993 *Workplace atmospheres. Size fraction definitions for measurement of airborne particles* British Standards Institution

14   *Lead and inorganic compounds of lead in air: Laboratory method using flame or electrothermal atomic absorption spectrometry* MDHS6/3 (Third edition) HSE Books 1998 ISBN 0 7176 1517 0

15   *Manual of analytical methods*, 4th edition (Department of Health and Human Services (NIOSH) Publication No 94-113), methods 2533 (tetraethyl lead) and 2524 (tetramethyl lead), USA

16   *Monitoring strategies for toxic substances* HSG173 HSE Books 1997 ISBN 0 7176 1411 5

17   *Lead and you: A guide to working safely with lead* Leaflet INDG308(rev1) HSE Books 1998 (single copy free or priced packs of 15 ISBN 0 7176 1523 5)

18   BS 1710: 1984 *Specification for identification of pipelines and services* British Standards Institution

19   BS 3900: Part B3: 1983 *Methods of test for paints. Tests involving chemical examination of liquid paints and dried paint films. Determination of 'soluble' lead in the solid matter in liquid paints: Methods for use in conjunction with the Control of Lead at Work Regulations* British Standards Institution

20   BS EN 13205: 2002 *Workplace atmospheres. Assessment of performance of instruments for measurement of airborne particle concentrations* British Standards Institution

21   BS EN 1232: 1997 *Workplace atmospheres. Pumps for personal sampling of chemical agents. Requirements and test methods* British Standards Institution

22   J R Davis and S L Andelman 'Urinary delta-aminolevulinic acid levels in lead poisoning. A modified method for the rapid determination of urinary delta-aminolevulinic acid using disposable ion-exchange chromatographic columns' *Arch. Environ. Health,* **15** (1967), 53-9

23   W E Blumberg, J Eisinger, A A Lamola and D M Zuckerman 'Zinc protoporphyrin level in blood determination by a portable haematofluorimeter. A screening device for lead poisoning' *J. Lab. Clin. Med.* **89** (1977), 712-23

24   Council Directive 77/312/EEC of 29 March 1977 on biological screening of the population for lead. OJ No L105 (1977) p.10 (Annex III)

25   A Berlin and K H Schaller 'European standardised method for the determination of delta-aminolevulinic acid dehydratase activity in blood' 3. *Klin. Chem. Klin. Biochem.* **12** (1974), 389-90

26   *A guide to the Reporting of Injuries, Diseases and Dangerous Occurrences Regulations 1995* L73 (Second edition) HSE Books 1999 ISBN 0 7176 2431 5

For details of how to obtain HSE Books publications, see inside back cover.

British Standards are available from BSI Customer Services, 389 Chiswick High Road, London W4 4AL Tel: 020 8996 9001 Fax: 020 8996 7001 Website: www.bsi-global.com.

The Stationery Office (formerly HMSO) publications are available from The Publications Centre, PO Box 276, London SW8 5DT Tel: 0870 600 5522 Fax: 0870 600 5533 Website: www.tso.co.uk (They are also available from bookshops.)

Further information

RIDDOR

Under the Reporting of Injuries, Diseases and Dangerous Occurrences Regulations 1995, all accidents, diseases and dangerous occurrences can now be reported to a central point.

Send RIDDOR reports to:
Incident Contact Centre,
Caerphilly Business Park,
Caerphilly CF83 3GG.
Tel: 0845 300 9923
Fax: 0845 300 9924
website: www.riddor.gov.uk
e-mail: riddor@natbrit.com

Printed and published by the Health and Safety Executive   C100   11/02